Research Series on the Chinese Dream and China's Development Path

Drawing on a large body of empirical studies done over the last two decades, the *Research Series on the Chinese Dream and China's Development Path* seeks to provide its readers with in-depth analyses of the past and present, and forecasts for the future course of China's development. Thanks to the adoption of Socialism with Chinese characteristics, and the implementation of comprehensive reform and opening, China has made tremendous achievements in areas such as political reform, economic development, and social construction, and is making great strides towards the realization of the Chinese dream of national rejuvenation. In addition to presenting a detailed account of many of these achievements, the authors also discuss what lessons other countries can learn from China's experience. This series will be an invaluable companion to every researcher who is trying to gain a deeper understanding of the development model, path and experience unique to China.

More information about this series at http://www.springer.com/series/13571

Houkai Wei

Urbanization in China

The Path to Harmony and Prosperity

SSAP

Springer

Houkai Wei
Chinese Academy of Social Sciences
Beijing, China

Sponsored by the Chinese Fund for the Humanities and Social Sciences

ISSN 2363-6866 ISSN 2363-6874 (electronic)
Research Series on the Chinese Dream and China's Development Path
ISBN 978-981-13-1407-0 ISBN 978-981-13-1408-7 (eBook)
https://doi.org/10.1007/978-981-13-1408-7

Jointly published with Social Sciences Academic Press, Beijing, China

The printed edition is not for sale in the Mainland of China. Customers from the Mainland of China please order the print book from: Social Sciences Academic Press.

Library of Congress Control Number: 2018947492

This Springer imprint is published by the registered company Springer Nature Singapore Pte Ltd.
The registered company address is: 152 Beach Road, #21-01/04 Gateway East, Singapore 189721, Singapore

Series Preface

Since China's reform and opening began in 1978, the country has come a long way on the path of Socialism with Chinese Characteristics, under the leadership of the Communist Party of China. Over thirty years of reform, efforts and sustained spectacular economic growth have turned China into the world's second-largest economy and wrought many profound changes in the Chinese society. These historically significant developments have been garnering increasing attention from scholars, governments, and the general public alike around the world since the 1990s, when the newest wave of China studies began to gather steam. Some of the hottest topics have included the so-called China miracle, Chinese phenomenon, Chinese experience, Chinese path, and the Chinese model. Homegrown researchers have soon followed suit. Already hugely productive, this vibrant field is putting out a large number of books each year, with Social Sciences Academic Press alone having published hundreds of titles on a wide range of subjects.

Because most of these books have been written and published in Chinese; however, readership has been limited outside China—even among many who study China—for whom English is still the lingua franca. This language barrier has been an impediment to efforts by academia, business communities, and policy-makers in other countries to form a thorough understanding of contemporary China, of what is distinct about China's past and present may mean not only for her future but also for the future of the world. The need to remove such an impediment is both real and urgent, and the *Research Series on the Chinese Dream and China's Development Path* is my answer to the call.

This series features some of the most notable achievements from the last 20 years by scholars in China in a variety of research topics related to reform and opening. They include both theoretical explorations and empirical studies and cover economy, society, politics, law, culture, and ecology; the six areas in which reform and opening policies have had the deepest impact and farthest-reaching consequences for the country. Authors for the series have also tried to articulate their visions of the "Chinese Dream" and how the country can realize it in these fields and beyond.

All of the editors and authors for the *Research Series on the Chinese Dream and China's Development Path* are both longtime students of reform and opening and recognized authorities in their respective academic fields. Their credentials and expertise lend credibility to these books, each of which has been subjected to a rigorous peer review process for inclusion in the series. As part of the Reform and Development Program under the State Administration of Press, Publication, Radio, Film, and Television of the People's Republic of China, the series is published by Springer, a Germany-based academic publisher of international repute, and distributed overseas. I am confident that it will help fill a lacuna in studies of China in the era of reform and opening.

Xie Shouguang

Preface

China has created a series of economic miracles since reform and opening up began. These miracles are primarily embodied in three aspects. First, China has enjoyed high economic growth for more than three decades, with an annual gross domestic product (GDP) growth rate of 9.8% between 1979 and 2013. Second, China has managed to lift about 500 million Chinese citizens out of poverty over the past 35 years, a significant contribution to global efforts against poverty. Third, urbanization in China is occurring at an unprecedented rate; China's urban population as a percentage of its total population, also known as the urbanization rate, rose from 17.92% in 1978 to 53.73% in 2013, a net increase of 559 million urban residents. Sustained high economic growth, poverty reduction and large-scale rapid urbanization constitute the three cornerstones of China's economic miracle and serve as the three pillars of the Chinese model of economic success.

As demonstrated by the Chinese experience, urbanization is the main driving force for sustained high economic growth and poverty reduction in rural areas. It is also an important approach to realizing the "Chinese Dream," that is, the revival of China. A significant part of the Chinese Dream is to assure equal rights between urban and rural residents and freedom of human migration, so that farmers are able to leave the countryside and agricultural work—on a voluntary basis—to pursue work in cities in non-agricultural sectors and enjoy the benefits of modern urban life. Achieving these goals will facilitate China's transition from a largely rural society into an urban one. This is also a long-held dream of Chinese farmers, especially rural migrant workers.

Migrating into cities and enjoying urban life are also a dream of farmers around the world. The global urban population increased from 745 million in 1950 to 3.8 billion in 2014. This corresponds to an increase in urbanization rate from 29.4 to 53.6%. By the end of 2014, the average urbanization rate in developed countries reached 78.0%, with some of the highest urbanization rates in countries that have essentially completed the process of urbanization, such as Japan (93%), Australia (89.3%), Canada (81.6%), the USA (81.4%), the UK (82.3%), and France (79.3%).

In contrast, the average urbanization rate in underdeveloped regions was 48.4%,[1] with fast urbanization still taking place. During the process of urbanization, however, all developing countries have faced severe problems, such as urban poverty, a large number of slums, and so-called urban diseases.

Moving into cities to enjoy a better, modern urban life has long been a dream of Chinese farmers. Before China began reform and opening up, this dream could only be realized through a few limited means due to the restrictive *hukou* system or household registration system. Such means included being hired by a city-based enterprise, becoming a government official, or attending university. After reform and opening, China saw a rapid increase in the number of rural migrant workers who moved into cities to work or do business. In 1983, there were about two million non-local rural migrant workers across China. This number has steadily risen over the years to 30 million people in 1989, 70 million people in 1995, 104.70 million people in 2002, and 166.10 million people as of 2013. The total number of both local and non-local rural migrant workers has already reached 268.94 million people in China.[2]

Most of these rural migrant workers work and live in cities and, to a certain extent, have realized the dream of becoming an urban resident. However, due to existing institutional limitations, they do not receive the same treatments as urban residents in terms of political rights, employment, childhood education, medical/health care, and social security. Moreover, given the low percentage of those who have become registered city residents, rural migrant workers are still far from integrated into urban life. Estimates show that by the end of 2012, the percentage of registered city residents among rural migrant workers in China was only about 40%, creating a huge gap between them and urban residents.[3] Accordingly, China still has a long way to go in terms of helping rural migrant workers achieve the dream of becoming a registered city resident.

This dream is not just about allowing Chinese farmers to move into cities, but more importantly, enabling them to establish a stable, good life so that they become integrated into urban society and contribute to social harmony. This approach represents high-quality healthy urbanization. There are three main indicators for qualifying successful urbanization in China: First is the achievement of an 80% urbanization rate (80% of the country's total population living in towns or cities), which would mark China's transition to an advanced stage of urban society. Second is freedom of human migration, in which rural migrants into cities have equal access to political rights and social welfare, or in other words, an end to discrimination against rural migrants. The third qualifier is the creation of eco-friendly, beautiful cities where there are no slums, and citizens are fully employed, affluent, and happy.

[1] This is a mid-year figure. Please refer to United Nations, Department of Economic and Social Affairs, Population Division (2014). World Urbanization Prospects: The 2014 Revision, CD-ROM Edition.

[2] The National Bureau of Statistics of China (NBS), "Migrant Workers Monitoring Survey Report 2013," May 12, 2014.

[3] Wei Houkai, Su Hongjian, and Li Fengtao, "Report on the Status of the Hukou urbanization of Rural Migrant Workers," *China Economic Weekly*, March 10, 2014.

Urbanization is an important approach to realizing the Chinese farmer's dream of becoming a registered city resident. In the future, China's urbanization will continue at a fast pace for about another two decades. It is expected that the urbanization rate in China will reach about 60% by the end of 2020, which corresponds to an urban population of about 840 million and China's transition to an intermediate stage of urban society. By the end of 2030, the urbanization rate in China is forecast to rise to about 68%, corresponding to an urban population that exceeds 950 million. Around 2033, it is expected that China will pass the 70% urbanization rate threshold to transition from an intermediate urbanization stage represented by fast urbanization to a late stage characterized by a slower but steady urbanization rate. By the end of 2040, the urbanization rate in China is anticipated to reach about 75%, which corresponds to an urban population of about 1.3 billion and represents an advanced stage of urban society. By the end of 2050, the urbanization rate in China will exceed 80%, close to the peak urbanization rate of 85%, when the urbanization rate will start to become stable. This means that, within three or four decades, the Chinese farmer's dream of becoming a registered city resident will be fulfilled—and about 80% of all Chinese will live in eco-friendly towns or cities where they are fully employed, affluent, and happy.

As a country with a large population, China faces significant rural–urban differences and limited availability of resources like energy, minerals, water, and land. This means that China should not conduct urbanization in the same way as any other single country or region, but should take an all-new approach that takes into account domestic conditions. A new China-specific approach to urbanization is an important part of China's efforts to develop into a unique socialist country; it is also fundamental to building a harmonious and prosperous society and reviving the nation. Global experience shows that sustainable overall prosperity and the revival of the nation will be unattainable if China does not achieve high-quality, healthy urbanization in a well-coordinated manner. Nor will this goal be achieved if hundreds of millions of farmers remain in the countryside and if rural migrants to cities cannot integrate into urban society. A new, China-specific approach to urbanization is necessary for China to build a harmonious and prosperous society and to realize the Chinese Dream.

It is around the aforementioned thread that this book is structured. Chapter 1 provides a history and basic characteristics of urbanization in China, assesses the current problems and challenges China faces, and reviews China's previous urbanization strategies. Chapter 2 presents the concept of "China-Model urbanization," focusing on the scientific basis, various implications, and basic characteristics of the new China-Model approach, as well as its combination with the concept of "new urbanization." Chapter 3 analyzes the changes in China's urbanization strategy, with a focus on relevant trends through 2050, as well as the near- and mid-term objectives and main tasks in this area. Chapters 4 and 5 explain how to conduct scientifically sound China-Model, new urbanization, including redefining the distribution of city sizes and optimizing the spatial distribution of cities. Chapter 6 explores how the administrative levels of Chinese cities relate to their size growth and analyzes how the bias toward administrative centers affects urban size growth. Chapter 7 explores how the spatial expansion of Chinese cities affects economic efficiency, carbon emissions, spatial structure, and other factors. Chapter 8 explains how to comprehensively improve urban development and management by such means as modernizing urban development and management, changing the dualistic spatial structure, addressing urban diseases, and building new types of cities. Chapter 9 explores institutional innovation for urbanization in such fields as the *hukou* system, land, employment, social security, and administrative division management.

This book was written with the assistance of Dr. Guan Xingliang and Dr. Su Hongjian, who participated in drafting and information collation for Chaps. 1, 2, 5, 7 and 8. I thank them for their input.

Beijing, China Houkai Wei
August 2014

Contents

Chapter 1
Urbanization in China: A Historical Review

With regard to urbanization, China has made remarkable achievements since it began reform and opening up. Such achievements include a notable increase in the urbanization rate, the growing dominance of urban economies, the growing role of central cities, significant improvements in urban infrastructures, public services and standard of living, fast-growing urban agglomerations, and the increasing ripple effects of urbanization. Nonetheless, China's traditional manner of conducting urbanization in recent years, along with the fast rate of urbanization, has led to serious problems, such as high consumption of resources, energy and land and huge amounts of waste; urban sprawl in towns and cities; developmental imbalances between rural and urban areas; and non-inclusiveness. Urban development has been taking place without considering local features, and there has been a general lack of management during urbanization. These problems indicate that China remains far from achieving high-quality, healthy urbanization. The traditional approach to urbanization must be replaced by a new China-Model urbanization model based on the new concept of "scientifically-sound development."

1.1 Administrative Hierarchy in Chinese Towns and Cities

Urbanization is an inevitable trend of human society and an important mark of human civilization and modernization. The meaning of urbanization varies from discipline to discipline based on different perspectives and focuses. Urbanization generally refers to the process of sustained human migration into towns and cities, which then expand in size, resulting in a series of socioeconomic changes. Urbanization is essentially a description of specific changes in economic, social and spatial structures. As it relates to economic structure, urbanization is the process where agricultural activities are gradually replaced by non-agricultural ones, leading to an improved industry structure. With respect to social structure, urbanization is the process where rural residents gradually become urban residents, and urban cultures, lifestyles and

H. Wei, *Urbanization in China*, Research Series on the Chinese Dream and China's Development Path, https://doi.org/10.1007/978-981-13-1408-7_1

values spread to the countryside. With regard to spatial structure, urbanization is the process where factors of production and industrial activities aggregate in urban areas before once again becoming decentralized (Wei 2005). Urbanization typically has multiple effects; it not only stimulates investment, domestic demand, employment and economic growth, but is also able to promote the optimization of the industry structure, speed up human capital accumulation and improve the standard of living. Globally, the percentage of the urban population to the total population is a common measure of a nation's level of urbanization. However, since the definition of towns and cities varies from country to country, there are difficulties in comparing urbanization levels on an international basis. The complex concept of what comprises a "town and city" and changeable urban statistical metrics, in particular, make it very necessary to accurately understand the administrative hierarchy in Chinese towns and cities (Zhou 1989). It is therefore necessary to first discuss the Chinese standards for the establishment of towns and cities as well as their administrative hierarchy, before closely studying the process of urbanization in China.

In China, towns and cities refer to settlements that are dominated by people working in non-agricultural sectors and that house sizable industry and commerce. *Ci Hai*, an encyclopedic dictionary published in China, does not explain the term "towns and cities", although we can indirectly deduce its meaning from the dictionary's explanation of the concepts of "urban population" and "rural population"—"Towns and cities are settlements dominated by non-agricultural economies" (Liu 2005). According to the *Regulations on Statistically Dividing Rural and Urban Areas (Trial)* issued by the National Bureau of Statistics of China (NBS) in 1999, "Towns and cities refer to the ones specified herein on the basis of China's urban system and administrative division. Cities herein refer to urban areas of municipalities set up with approval from the State Council, or China's Cabinet, including districted and non-districted ones. Towns herein refer to areas of towns set up with approval from the competent government authorities, including those of towns as the seats of the people's governments and administrative offices at the county or higher levels and of the other towns." The organizational framework of Chinese towns and cities are shown in Fig. 1.1.

Globally, there are no uniform standards for defining towns and cities. Standards for setting up towns and cities vary from country to country and, within the same country, from region/stage to region/stage (Zhou 2007). There are standards based on the seats of administrative centers at a particular level (e.g., Mongolia), on the characteristics of towns and cities (e.g., Chile), on the population size of an area (e.g., Mexico and Iran), on a combination of the population size and density of an area (e.g., Sweden and Canada), on a combination of the population size and employment structure of an area (e.g., The Netherlands), and so forth. China has more complex standards for setting up towns and cities than in most other countries. These standards combine various factors such as population size and density, the city's functions and the developmental levels of non-agricultural sectors. Today, China sets up cities according to the *Notice of the State Council on Approving and Forwarding the Ministry of Civil Affairs' Report on Adjusting the Standards for Setting Up Cities* (Document GF [1993] No. 38), as shown in Tables 1.1 and 1.2.

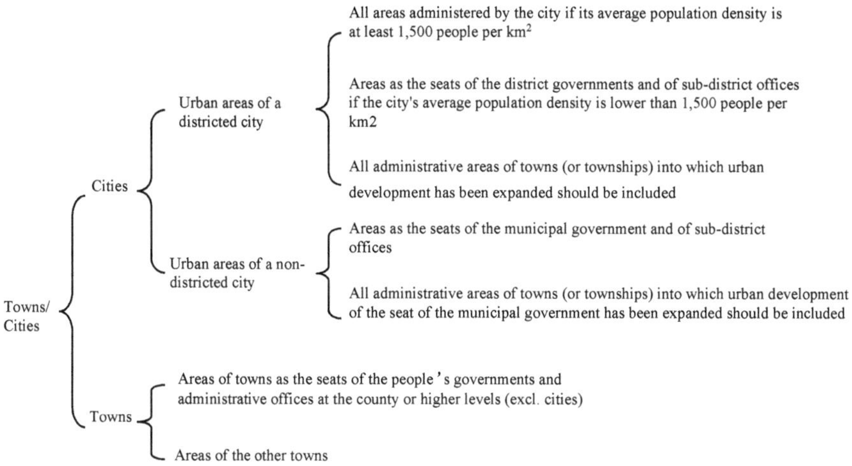

Fig. 1.1 Framework of Chinese towns and cities. *Source* The *Regulations on Statistically Dividing Rural and Urban Areas (Trial)* issued by the NBS in 1999

Specifically, the standards for setting up municipalities at the county level differ from those for setting up municipalities at the prefecture level, as the latter are more stringent than the former. China has changed its standards for setting up towns three times since the People's Republic of China was founded in 1949. The most recent change was made in 1984. According to the *Notice of the State Council on Approving and Forwarding the Ministry of Civil Affairs' Report on Adjusting the Standards for Setting Up Towns* (Document GF[1984] No. 165), "All the seats of local governments at the county level shall be set up as towns; a township which has a total population of fewer than 20,000 people and where the seat of the township government has a non-agricultural population of more than 2000 people, or a township which has a total population of at least 20,000 people and where the seat of the township government has a non-agricultural population which accounts for at least 10% of this township's total population, may be set up as a town; areas dominated by minority ethnic groups, thinly populated remote areas, mountainous areas, small industrial and mining areas, small ports, tourism-centric areas and border areas each with a non-agricultural population of fewer than 2000 people, also may be set up as towns if necessary."

Chinese cities fall under different types based on population size. Under the former *Urban Planning Law of the People's Republic of China*, a city is a large city if the non-agricultural population in the urban and suburban areas is at least 500,000 people, a medium-sized city if that same population is between 200,000 and 500,000 people, and a small city if that population is smaller than 200,000 people. Meanwhile, given China's administrative system, Chinese towns and cities are also ranked at different administrative levels, including municipalities directly under the Central Government, and those at the sub-provincial, prefecture, sub-prefecture, county, and town levels. Chinese towns and cities have certain

Table 1.1 Chinese standards for setting up municipalities at the county level

	Population density (in person/km^2)		>400	100–400	<100
Seat of county government	Non-agricultural population (in 1000 people)		≥120	≥100	≥80
	Including: Population with urban *hukou* (in 1000 people)		≥80	≥70	≥60
	Tap water penetration (%)		≥65	≥60	≥55
	Road penetration (%)		≥60	≥55	≥50
	Good urban infrastructure and sewage system				
County	Non-agricultural population (in 1000 people)		≥150	≥120	≥100
	The percentage of non-agricultural population to total population (%)		≥30	≥25	≥20
	Industrial output value from enterprises above the township and village enterprise (TVE) level (billion yuan, according to the 1990 price level, the same below)		≥1.5	≥1.2	≥0.8
	The percentage of industrial output value from enterprises above the TVE level to the gross value of industrial and agricultural output (GVIAO) (%)		≥80	≥70	≥60
	GDP (in billion yuan)		≥1.0	≥0.8	≥0.6
	The percentage of the tertiary sector in GDP (%)		≥20	≥20	≥20
	Budgeted local revenue	Total (in million yuan)	≥60	≥50	≥40
		Average (in yuan/person)	≥100	≥80	≥60
		Be liable to turn over a certain percentage of revenue to the upper-level authority			

Source Document GF [1993] No. 38

Table 1.2 Criteria in China for setting up municipalities at the prefecture level

Non-agricultural population in urban areas	1000 people	≥250
Population with urban *hukou* at the seat of the municipal government	1000 people	≥200
GVIAO	Billion yuan	≥3.0
The percentage of industrial output value to GVIAO	%	≥80
GDP	Billion yuan	≥2.5
The percentage of the tertiary sector to GDP	%	≥35% and greater than that of the primary sector
Budgeted local revenue	Billion yuan	≥0.2

Source Document GF [1993] No. 38

powers with respect to resource allocation and administration depending on their administrative levels, as defined under certain historical circumstances. Under the *Constitution of the People's Republic of China*, administrative areas in China consist of three basic levels—township, county and province. China began rolling out the city-administered (or city-led) county system in combination with overall economic system reforms shortly after it began reform and opening up. This move was aimed at addressing the problems of the urban-rural and sectorial divide, enabling ripple effects from economically developed cities, and establishing city-centric economic zones. Accordingly, in 1982 the Constitution was amended from the 1954 version that read "Municipalities directly under the Central Government and other larger municipalities shall be divided into districts and counties" to a new provision stating that "Municipalities directly under the Central Government and other larger municipalities shall be divided into counties." This amendment legalized city-administered counties. The city-administered county system has fueled the rise of municipalities at the prefecture level, which in turn have become a new administrative level between counties and provinces (Wang 2013). After more than three decades of development, a complete set of administrative level towns and cities have been established in China on the basis of the city-administered county system.

Municipalities directly under the Central Government refer to cities directly administered by the Central Government. As administrative units at the provincial level, such municipalities have the same status as provinces and autonomous regions. According to Article 30 of the Constitution of the People's Republic of China, "the country is divided into provinces, autonomous regions and municipalities directly under the Central Government." Municipalities directly under the Central Government are identical to other larger municipalities in that they generally administer districts and counties. By the end of 2013, there were four municipalities directly under the Central Government in China, including Beijing, Tianjin, Shanghai and Chongqing.

Municipalities at the sub-provincial level first appeared in 1994. Their predecessor was the so-called "cities specifically designated in the state plan", which were administered by administrative areas at the provincial level. Shortly after reform and opening up, China began designating particular cities in its state plans, granting them the same socioeconomic authority as provinces. Once again, the purpose of this move was to address the problems of the urban-rural and sectorial divide, as well as explore approaches to establish large-city-centric economic zones and develop the economy through independence from politics. In February 1983, the State Council identified the city of Chongqing as a "designated city" in its state plan. Later on, the cities of Wuhan, Shenyang, Dalian, Harbin, Xi'an, Guangzhou, Chengdu, Qingdao, Ningbo, Xiamen, Shenzhen, Nanjing and Changchun also became specifically designated cities in the state plan. By the end of 1993, fourteen Chinese cities were specifically designated in the state plan, including Shenzhen, Ningbo, Dalian and Xiamen as non-provincial capitals. In 1994, the Central Committee of the Communist Party of China (CPC) and the State Council consented to establish municipalities at the sub-provincial level to replace the former system in which provincial capitals were specifically "designated cities" in the state plan. This change was formalized in a document issued by the State Commission for Public Sector Reform (SCOPSR) on February 25, 1994. As a result, the former fourteen cities specifically designated in the state plan, plus Hangzhou and Jinan, were identified as municipalities at the sub-provincial level. In March 1997, Chongqing was raised to a municipality directly under the Central Government. Accordingly, the number of municipalities at the sub-provincial level was reduced to fifteen, including ten provincial capitals and five cities formerly designated in the state plan. From the perspective of administrative division, municipalities at the sub-provincial level also are municipalities at the prefecture level.

Municipalities at the prefecture level refer to municipalities that have the same administrative status as that of prefectures, autonomous prefectures and leagues. They are administrative areas at the prefecture level and are led by relevant administrative units at the provincial level. A city at the prefecture level typically administers districts, counties, autonomous counties, banners, autonomous banners and municipalities at the county level. Theoretically, municipalities at the prefecture level, including municipalities at the sub-provincial level, are quasi-administrative areas, with administrative status between that of counties and provinces, according to China's current administrative division system consisting of townships, counties and provinces. By the end of 2013, there were 286 municipalities at the prefecture level in China.

Municipalities at the sub-prefecture level refer to province-administered municipalities at the county level, most of which were raised from municipalities at the county level. With an administrative level between cities at the county and prefecture levels, municipalities at the sub-prefecture level are still non-districted cities.

Municipalities at the county level refer to cities that have the same administrative status as counties. Most municipalities at the county level were raised from counties during the industrialization and urbanization process that has accelerated since the

1980s. Such cities are at the same administrative division level as that of counties. By the end of 2013, there were 368 municipalities at the county level in China.

Towns refer to those set up with approval from the governments of relevant provinces, autonomous regions or municipalities directly under the Central Government. At the same administrative level as townships, towns are the lowest-level administrative units in China. By the end of 2013, there were 40,497 administrative areas at the township and town levels, including 20,117 towns, in China.

1.2 Stages of Urbanization in China

The process of urbanization since the founding of the People's Republic of China more than six decades ago can be roughly divided into the following stages: the oscillating stage (1950–1977), the steady advancement stage (1978–1995) and the acceleration stage (1996–present) (see Fig. 1.2). The stages and variations of this process have been affected by a combination of factors, including domestic and international environments, state policies and urbanization-relevant guidelines.

1.2.1 Oscillation (1950–1977)

Before China began reform and opening up, the process of urbanization in this country can be considered oscillating due to changes in state policies. This stage

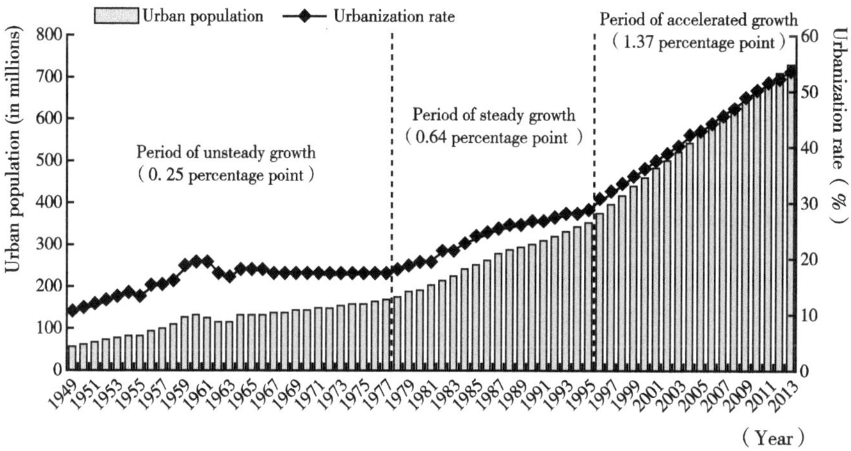

Fig. 1.2 Stages of urbanization in China. Source: *China Statistics Yearbook* (in previous years) and the *Statistical Communiqué of the People's Republic of China on the 2013 National Economic and Social Development*

can be divided into six sub-stages. Sub-stage 1 (1950–1953) was characterized by restoration and steady advancement, with the nationwide urbanization rate rising from 10.64% in 1949 to 13.31% in 1953, an annual average increase of 0.67 percentage points. Sub-stage 2 (1954–1955) was characterized by stagnation, with the urbanization rate staying around 13.5%. Sub-stage 3 (1956–1960) was characterized by a sharp increase amid the Great Leap Forward economic and social campaign, with the urbanization rate rising rapidly from 13.48% in 1955 to 19.75% in 1960, an annual average increase of 1.25 percentage points, which was beyond what the national economy could afford. Sub-stage 4 (1961–1963) was characterized by anti-urbanization, with the urbanization rate dropping by 2.91 percentage points within three years, an annual average decrease of 0.97 percentage points. Sub-stage 5 (1964) was characterized by a rebound, with the urbanization rate increasing rapidly by 1.53 percentage points. Sub-stage 6 (1965–1977) was characterized by stagnation and decline as a result of the Cultural Revolution; the nationwide urbanization rate during this period hovered at around 17.5%, and even showed a slight downward trend.

During this first stage, China's overall urbanization progressed relatively slowly and was characterized by multiple variations and uncertainties. The first three years showed recovery and steady development as the 1st Five-Year Plan period kicked off, the three-year Great Leap Forward campaign spiked urbanization as the national economy adjusted, and the Cultural Revolution caused a decline as a result of the relocation of a great number of enterprises from coastal regions to inland ones. The great ups and downs caused by a series of external factors constituted the primary challenge to urbanization in this stage (Zhou 2007; Liu 2011). Among these factors include China's weak socioeconomic basis, uncertainties in the domestic and international environments, blind copying of the pattern of the former Soviet Union, overemphasis on the development of heavy and chemical industries, an industrialization strategy that neglected the development of non-production sectors, reckless policies during the Great Leap Forward campaign, excessive pessimism about the international situation, excessive industrial projects across China, and neglect of the leading role of large cities (see Table 1.3). Additionally, natural disasters and a worsening relationship between China and the former Soviet Union were other significant factors. Under the combined effects of these internal and external factors did China develop the policies and measures that has set the tone for China's urbanization in the years to follow, such as the household registration system, restrictions on domestic human migration, the relocation of enterprises from coastal regions to inland regions, control over the size of large cities, the focus on the development of towns, and the suspension of urban planning (Zhou 2007).

1.2.2 Steady Progress (1978–1995)

China began shifting the focus of its attention toward socialist modernization after the end of the Cultural Revolution, especially after 1978. Great changes have since taken place in the political, economic and other fields, and urbanization has also

Table 1.3 Stages of urbanization in China: characteristics, achievements and problems

Stage	Characteristics	Achievements	Driving factors	Problems
Oscillation (1950–1977)	Great ups and downs; slow rate of growth	• The urbanization rate rose from 10.64% in 1949 to 17.55% in 1977 at an annual average increase of 0.25 percentage points • The urban population grew from 57.65 million people in 1949 to 166.69 million people in 1977 at an annual average net increase of 3.89 million people	• Eagerness for rapid success, which contributed to recklessness and great ups and downs. Great Leap Forward had far-reaching consequences on urbanization • Copying the pattern of the former Soviet Union, which led to overemphasis on the development of the heavy and chemical industries, and neglect of the development of non-production sectors • The Cultural Revolution, in which intellectuals and government officials were sent to labor camps, and young urban residents were sent to the countryside • Excessive pessimism about the international situation; a worsening relationship between China and the former Soviet Union, and so forth • The relocation of particular enterprises from coastal to inland regions and the guideline of “control the sizes of large cities while developing towns” • Natural disasters	• Heavy industry became absolutely dominant, while light industry and the tertiary sector were neglected, resulting in a severely unbalanced economic structure • The decentralization-oriented relocation of particular enterprises from coastal to inland regions, with factories being moved into valleys or even caves, causing economic inefficiency • Developmental limitations for large cities, and lack of driving forces for development in towns, leading to failure in achieving substantial urban development • A household registration system that restricted human migration became the root cause of the urban-rural divide • The suspension of urban planning and other measures had far-reaching consequences on urban development

(continued)

Table 1.3 (continued)

Stage	Characteristics	Achievements	Driving factors	Problems
Steady Advancement (1978–1995)	Steady advancement; on the right track	• The urbanization rate rose from 17.55% in 1977 to 29.04% in 1995 at an annual average increase of 0.64 percentage points • The urban population grew from 166.69 million people in 1977 to 351.74 million people in 1995 at an annual average net increase of 10.28 million people	• China shifted the focus of its attention toward socialist modernization after it began reform and opening up, with urbanization back on track • China shifted the focus of its industrialization strategy from heavy and chemical industries toward light industry represented by consumer goods, leading to a significant increase in jobs available in towns and cities • Recalibration of China's urbanization strategy that prioritized towns over large cities • A growing number of farmers taking non-agricultural jobs in their hometowns without moving into cities became the primary pattern of industrialization in China, characterized by decentralized industrialization	• The great passion for economic development in the early stage of China's reform and opening up brought about investment and economic overheating, and the resulting tightening policies hugely impacted the Chinese economy • An excessive amount of town development led to wastage of land and other resources, lack of economies of scale and environmental deterioration • Medium-sized and large cities were unable to play a desirable role in terms of economies of scale and agglomerative and ripple effects; towns and cities could only offer jobs to a limited portion of the workforce

(continued)

Table 1.3 (continued)

Stage	Characteristics	Achievements	Driving factors	Problems
Acceleration (1996–2013)	Faster urbanization	• The urbanization rate rose from 29.04% in 1995 to 53.73% in 2013 at an annual average increase of 1.37 percentage points • The urban population grew from 351.74 million in 1995 to 731.11 million people in 2013 at an annual average net increase of 21.08 million people	• With fast economic growth, China's industrialization and urbanization accelerated • China experienced continuous improvement in its industrial structure, with growing shares of the secondary and tertiary sectors and more jobs available in towns and cities • Renewed urbanization strategy focused on the development of large cities; emergence of an urban system characterized by well-coordinated development of all towns and cities; urban agglomerations began appearing as the main form of urbanization	• Urban economies grew rapidly with high consumption of resources and large amounts of waste, placing heavy pressure on resources and the environment • Towns and cities experienced urban sprawl • There was a widening gap between urban and rural areas • Non-inclusiveness in urbanization became a highly visible and significant problem • Urban development neglected local features, leading to uniform cities across China; there was a lack of management during urbanization

gotten back on track (Zhou 2007). First, at the beginning of this stage, the Chinese government announced an important national strategy centered on economic development. Second, it shifted the focus of China's industrialization strategy from heavy and chemical industries toward light industry represented by consumer goods, which lead to a significant increase in jobs available in towns and cities. Third, shaped by the restrictive household registration system and the emergence of Township and Village Enterprises (TVEs), the primary pattern of industrialization in China arose around large numbers of farmers taking non-agricultural jobs in their hometowns without moving into cities. With industrial development across the countryside, towns began to prosper and rapidly assumed a leading position among all Chinese towns and cities (Zhou and Sun 2012). Notwithstanding these positive developments, the changes during this stage had its limits and negative effects. The great passion for economic development in the early stage of China's reform and opening up led to investment and economic overheating, and the resulting tightening policies significantly impacted the Chinese economy (see Table 1.3). In addition, excessive development of towns, plus the ubiquity of industrial enterprises, resulted in wastage of land and other resources and led to severe environmental degradation. Because of government restrictions on city size limits, medium-sized and large cities were unable to play a desirable role in achieving economies of scale and the benefits of agglomerative and ripple effects. Chinese towns and cities also could only offer jobs to a limited portion of the workforce.

Generally, the period between the start of China's reform and opening in 1978–1995 was an era of steadily advancing urbanization in China. With the deepening of reform and opening up as well as rapid industrialization in this period, China saw a steadily increasing urbanization rate, which rose from 17.55% in 1977 to 29.04% in 1995, or 11.49 percentage points within 18 years, at an annual average increase of 0.64 percentage points. More specifically, urbanization in China occurred faster in the period between 1978 and 1987 with an annual average increase of 0.78 percentage points. In comparison, urbanization in China slowed down between 1988 and 1995 with an annual average increase of 0.47 percentage points, mainly due to rectifying policies issued in 1989. In brief, while urbanization steadily advanced during this stage, overall urbanization progress lagged far behind industrialization progress because of the faster rate of the latter combined with the negative effects of perennial problems.

1.2.3 Acceleration (1996–Present)

With further deepening of reform and opening up as well as steady economic development, China has enjoyed a faster rate of industrialization and urbanization since 1996 (see Table 1.3). China's urbanization rate surpassed the 30% mark in 1996, reaching 30.48%. China's per capita GDP topped $1000 USD in 2001. Both of these indicators reveal China's entrance to a stage of accelerated industrialization and urbanization. Also, in contrast to the pattern of decentralized industrialization

Table 1.4 Urbanization rate in China by period

Period (Five-Year Plan)	Year	Annual average net increase of urban population (in million people)	Annual average increase of urbanization rate (%)
6	1981–1985	11.91	0.86
7	1986–1990	10.20	0.54
8	1991–1995	9.96	0.53
9	1996–2000	21.46	1.44
10	2001–2005	20.61	1.35
11	2006–2010	21.53	1.39
12	2011–2013	20.44	1.26

Source: *China Statistics Yearbook* (in previous years) and the *Statistical Communiqué of the People's Republic of China on the 2013 National Economic and Social Development*

in the countryside and the approach to urbanization in the previous stage, both the government and academic sectors in the acceleration stage paid increasing attention to large cities with respect to economies of scale as well as agglomerative and ripple effects. In June 2000, the Central Committee of the CPC and the State Council jointly issued the *Opinions on Promoting the Healthy Development of Towns*, noting that "It is already the right time to accelerate urbanization." China's 10th and 11th Five-Year Plans specifically promoted urbanization as a national strategy, stating:

> It is necessary to take an approach to diversified urbanization that suits the conditions in China and that is characterized by well-coordinated development of all towns and cities. It is also necessary to develop towns with a focus on selected towns, actively develop small and medium-sized cities, improve the functions of regionally central cities, and allow large cities to play the leading role. We should take urban agglomerations as the main form of urbanization.

Thanks to a favorable macroeconomic environment and an aggressive urbanization strategy, urbanization in China has been accelerating since 1996. On average, the urbanization rate has been increasing by 1.37 percentage points per year over the past 18 years, which is much higher than the 0.25 percentage points growth in the 1950–1977 period and the 0.64 percentage points growth in the 1978–1995 period. Specifically, China's urban population grew from 562.12 million people in 2005 to 669.78 million people in the 10th Five-Year Plan period, a net increase of 107.66 million people, or an annual average increase of 21.53 million people. In the meantime, the urbanization rate rose from 42.99 to 49.95% at an annual average rate of 1.39 percentage points. By comparison, the annual average net increase of China's urban population and the urbanization rate in the 11th Five-Year Plan period were almost equal to that of the 9th and 10th Five-Year Plan periods combined, but still much higher than in the 7th and 8th Five-Year Plan periods (see Table 1.4).

Between 2011 and 2013, the annual average net increase of China's urban population was 20.44 million people, a growth in the urbanization rate of 1.26 percentage points per year on average, which is slightly lower than that of the 9th, 10th and 11th

Five-Year Plan periods. This indicates that urbanization in China is showing signs of deceleration in the 12th Five-Year Plan period, following the rapid development over the previous fifteen years. It is expected that urbanization in China will gradually slow down in the foreseeable future and enter a transformational stage in which the rate and quality of urbanization is viewed as equally important.

1.3 Basic Characteristics of Urbanization in China

With fast economic growth and industrialization since China began reform and opening up, urbanization in China has been progressing at an increasingly fast rate. Urban economies have become increasingly dominant thanks to the growing effects of urbanization. It took China only 30 years to achieve what it took the United Kingdom about a century to in terms of increases in urbanization rate, creating another "Chinese miracle." Urbanization in China has five chief characteristics when compared with the experience of other countries.

1.3.1 Low Starting Point, Large Scale and Fast Rate

Urbanization in China started from a low base. The urbanization rate in China was merely 19.4% in 1980, which was 20.0 percentage points lower than the global average and 10.1 percentage points lower than the average level of underdeveloped countries and regions. China's urban population has grown from 172 million people in 1978 to 731 million people in 2013. The urbanization rate in China during this period has risen from 17.92 to 53.73%, an annual average increase of 1.02 percentage points. This means that China's urban population grew by an average net increase of 15.97 million people every year since 1978. In particular, since the 9th Five-Year Plan period (starting in 1996), the urbanization rate in China has had an annual average increase of at least 1.25 percentage points in each period, with its urban population showing a consistent annual average net increase of over 20 million people. Such a rate and scale have been rarely seen anywhere else in the world. Between 1981 and 2010, the global urbanization rate grew by only 0.41 percentage points per year on average. More specifically, the urbanization rate grew at an annual average rate of 0.23 percentage points in developed regions and 0.56 percentage points in underdeveloped regions. This is in stark contrast to China's annual average urbanization rate growth rate of 1.00 percentage points, which is much higher than the global average growth rate (see Fig. 1.3). It took more than 50 years for the global urbanization rate to rise from 30 to 50%, whereas in China it took only 15 years.

1.3.2 Regional Disparity in Urbanization Rate

Notable regions disparity exists in China in terms of level of industrialization and economic development, and in urbanization rate. Generally, the eastern and northeastern regions have higher urbanization rates than central and western regions. Urbanization in the eastern regions started accelerating earlier than in the central and western regions, although urbanization in the northeastern regions has been progressing at a slower rate. By the end of 2012, the urbanization rates in the eastern and northeastern regions had reached 61.86 and 59.60% respectively, compared with 47.19 and 44.74% in the central and western regions respectively. The western regions were 17.12 and 14.86 percentage points lower than the eastern and northeastern regions respectively (see Table 1.5). During the period between 1978 and 2010, the urbanization rate in eastern regions grew by 1.37 percentage points per year on average, compared with a growth rate of 0.92, 0.86 and only 0.64 percentage points in the central, western and northeastern regions respectively. In this period, the urbanization rate gap between the eastern and western regions widened from 1.95 to 18.27 percentage points; and the gap between the eastern and central regions widened from 1.65 to 16.12 percentage points. Since 2010, however, growth in urbanization in the eastern regions has seen slowing down, whereas urbanization in the central and western regions is still accelerating. Now, with a higher urbanization rate than in the eastern regions, western regions are seeing a narrowing gap.

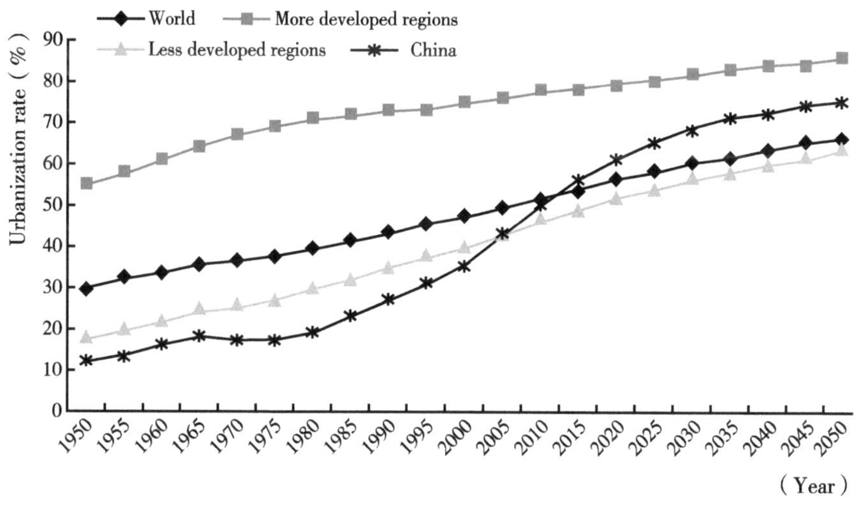

Fig. 1.3 Urbanization process: China versus the rest of the world. *Note* The figures of 2010 and earlier are actual figures; figures after 2010 are estimates

Table 1.5 Changes in urbanization rate by region, 1978–2012

Year	Urbanization rate (%)				Period	Annual average growth (percentage points)			
	Eastern	Northeastern	Central	Western		Eastern	Northeastern	Central	Western
1978	15.73	37.01	14.08	13.78	-	-	-	-	-
1980	17.91	39.08	15.23	15.76	1978–1980	1.09	1.04	0.58	0.99
1985	28.00	45.98	19.57	20.16	1981–1985	2.02	1.38	0.87	0.88
1990	32.44	48.69	22.32	21.96	1986–1990	0.89	0.54	0.55	0.36
1995	39.29	51.37	26.99	25.97	1991–1995	1.37	0.54	0.93	0.80
2000	44.61	52.26	29.82	28.68	1996–2000	1.06	0.18	0.57	0.54
2005	51.78	55.15	36.55	35.36	2001–2005	1.43	0.58	1.35	1.34
2010	59.70	57.62	43.58	41.43	2006–2010	1.59	0.49	1.41	1.21
2012	61.86	59.60	47.19	44.74	2011–2012	1.08	0.99	1.81	1.66

Note The eastern regions include the provinces of Hebei, Shandong, Jiangsu, Zhejiang, Fujian, Guangdong and Hainan as well as the cities of Beijing, Shanghai and Tianjin; the northeastern regions include the provinces of Liaoning, Jilin and Heilongjiang; the central regions include the provinces of Shanxi, Henan, Anhui, Hubei, Hunan and Jiangxi; the western regions include the autonomous regions of Inner Mongolia, Guangxi, Xinjiang, Ningxia and Tibet, the provinces of Shaanxi, Gansu, Qinghai, Sichuan, Guizhou and Yunnan, and the city of Chongqing

Source Wei (2014) and *China Statistics Yearbook* (2013)

1.3.3 The Rise of Urban Agglomeration as an Urbanization Model

Over the period covered by the past three Five-Year Plans, the Chinese government has gradually targeted urban agglomeration as the lynchpin of its urbanization plan. In the *Outline of the 10th Five-Year Plan,* the Chinese government stated: "… grasp opportunities for implementing the urbanization strategy." In the *Outline of the 11th Five-Year Plan* this evolved to: "… take urban agglomeration as the main urbanization model:" In the current *Outline of the 12th Five-Year Plan*, China's focus on urban agglomeration as a urbanization strategy has been further refined as: "gradually develop more internationally competitive urban agglomerations in the eastern regions while developing urban agglomerations in parts of the central and western regions where certain conditions and requirements are satisfied." Under relevant state policies, a number of urban agglomerations of different sizes and developmental levels have emerged across China (see Fig. 1.4). Typical examples include the Yangtze River Delta, Pearl River Delta, Beijing-Tianjin-Hebei, Shandong Peninsula, Central and Southern Liaoning, West Side of the Strait, Harbin-Daqing-Changchun, Central China, Changsha-Zhuzhou-Xiangtan, Wuhan and Chengdu-Chongqing urban agglomerations. As entirely new Chinese regional units that participate in global

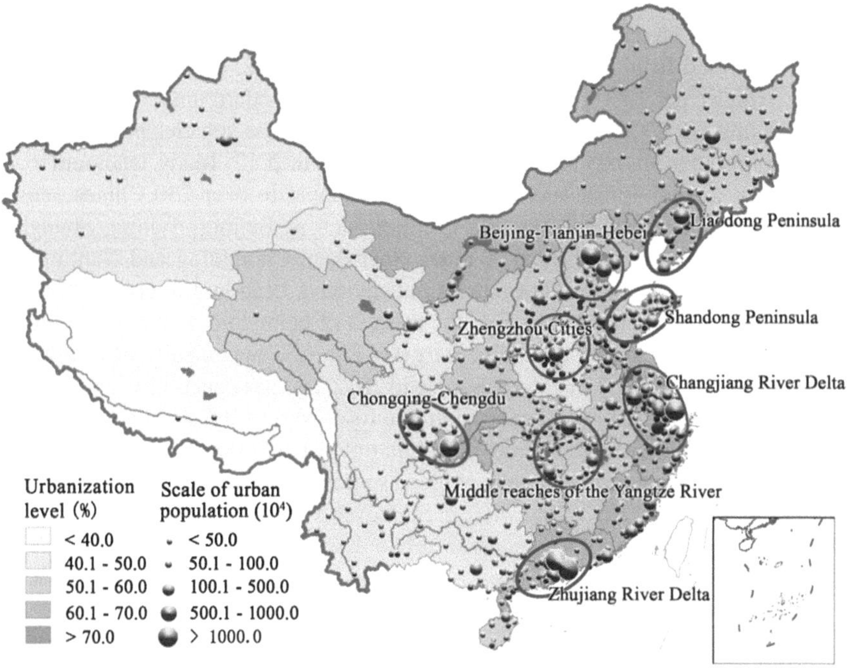

Fig. 1.4 Spatial pattern of urban agglomerations of China in 2015. *Note* The data comes from China Urban Construction Statistical Yearbook

competition and international division of labor, these urban agglomerations have become regions that lead and support China's fast economic growth, that is, they form the backbone of the Chinese economy (Yao et al. 2006; Fang et al. 2011). With the rapid emergence of urban agglomerations, China has entered a new era of competition between these new urban entities, which have assumed the lead role in driving China's urbanization and in receiving new members of the urban population. According to a study by Fang et al. (2011), there are 15 mature urban agglomerations in China and eight other still-developing agglomerations; these 23 urban agglomerations accounted for 46.7% of all Chinese towns and cities, 51.4% of China's urban population and 78.8% of its GDP in 2007. Their economic and population densities were 3.63 and 2.26 times higher than the national averages respectively. Specifically, the Yangtze River Delta, Pearl River Delta, Beijing-Tianjin-Hebei, Chengdu-Chongqing, Shandong Peninsula and Central and Southern Liaoning urban agglomerations accounted for 31% of China's urban population.

1.3.4 Rapid Development of Urban Infrastructure

Faster urbanization has fueled economic growth, urban industry structure improvements, and urban development, leading to better urban infrastructure and public services, as well as driving huge amounts of wealth for the Chinese (Fang 2009). Urban transport, in particular, has improved significantly; the average road area per capita increased from 1.8 m^2 in 1981 to 14.4 m^2 in 2012, the average number of public vehicles per 1000 people increased from 0.25 standard units in 1986 to 1.11 standard units in 2008, and the average number of private vehicles per 100 households increased from 0.03 units in 1990 to 21.5 units in 2012. Today, lanes and roads exclusively for public transport are already available in over 230 Chinese cities. Rail transport has become available in cities such as Beijing, Tianjin, Shanghai, Guangzhou, Changchun, Dalian, Wuhan, Shenzhen, Chongqing and Nanjing. By the end of 2012, a total of 2006 km of railway have been laid in sixteen Chinese cities, with an additional 2060 km in the process of being laid. Urban public utility penetration has also increased significantly (see Fig. 1.5). Specifically, tap water penetration rose from 53.7% in 1984 to 97.2% in 2012, and gas penetration from 11.6 to 93.2%; the sewage treatment penetration rate rose from 14.9% in 1991 to 87.3% in 2012, although the average number of public toilets per 1000 people stayed around 0.3 units. In addition, the percentage of green space in built-up areas rose from 19.1% in 1996 to 35.7% in 2012. Likewise, green coverage in such areas rose from 24.4 to 39.6%. These great achievements in urban development have led to continuous improvements in the overall competitiveness and functions of cities while providing a rising standard of living for urban residents.

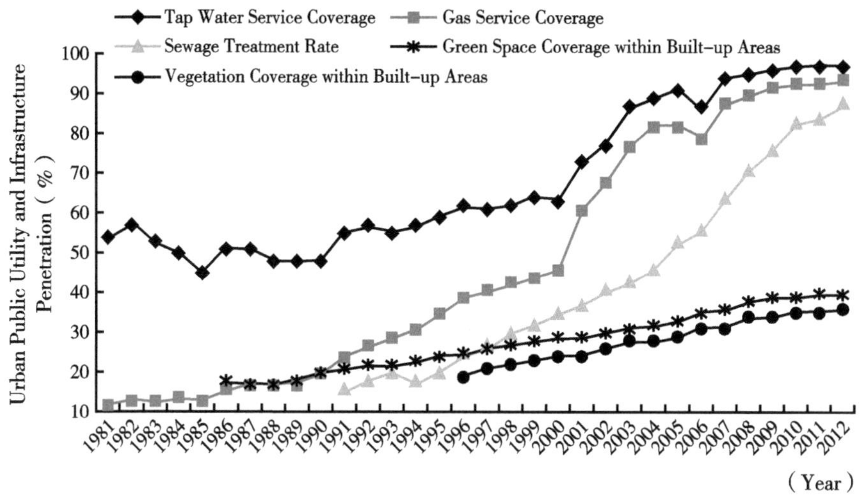

Fig. 1.5 Urban public utility and Infrastructure development in China by year. *Note* The tap water and gas penetration indicators since 2006 are calculated with the population in urban areas, including permanent and temporary residents Source: *China Urban Construction Statistical Yearbook* (2012)

1.3.5 Urbanization as an Important Engine for Economic Growth

Urbanization is the product of socioeconomic development and an important driving force for economic growth and structural change. China's fast economic growth since it began reform and opening up is closely tied with rapid urbanization, which has brought strong investor and consumer demand and thus has fueled economic growth. With respect to investor demand, an increase in urban population brings investor demand in various sectors such as urban infrastructure, public service facilities and real estate development. It is generally believed that for every anew member of the urban population, about 100,000 yuan investment is required. Accordingly, the total investment required to accommodate China's growing urban population has been over 2 trillion yuan per year since the 9th Five-Year Plan period. Such investment is an important force for economic growth. Every additional percentage point in urbanization rate would translate into a 3.7 percentage points increase in investment in the period from 2001 to 2011, according to estimates from the NBS. Investment growth triggered by urbanization can not only reduce pressure caused by overcapacity in the iron and steel, cement and other industries, but can also create a great deal of jobs for new members of the urban population (Liang 2012). With respect to consumer demand, the migration of a large part of the rural workforce from the countryside into towns and cities has significantly increased consumer demand. Every additional percentage point in China's urbanization rate indicates the migration of over 10 million farmers into towns and cities. Since urban residents spend much more money than rural residents (2.8 times more in 2012, for example), this migration will lead to a visible increase in end consumption. China's fast urbanization, combined with the movement of a large part of the rural workforce from the primary sector into the secondary and tertiary sectors, has greatly promoted China's economic growth. According to relevant studies, urbanization has contributed 5.0 percentage points to China's economic growth (Wang and Xia 2000); the inter-sectorial workforce movement has contributed about 16–20% to China's GDP growth (World Bank Group 1997; Cai and Wang 1999). In the meantime, urbanization in China has sped up human capital accumulation and promoted technological innovation, productivity, and improvements to the industrial structure (Shi 2012; Qiu 2013; Wei 2014). Furthermore, fast urbanization in China has facilitated improvements in basic public services and the narrowing of the urban-rural gap [China Development Research Foundation (CDRF) 2010].

1.4 Problems and Challenges

Despite its many achievements in urbanization over the past years, China has followed a traditional path to urbanization since it began reform and opening up, and today faces a series of severe problems that need to be addressed as soon as possible. Such problems include: prioritizing "land urbanization" over "population" urbanization, the latter of which seeks to improve the quality of life for residents; neglecting the allocative efficiency of urban resources and excessively consuming land, energy, water and other resources; the lack of coordination among production, people's livelihood and the environment; the urban-rural divide; the lack of local features in urban development, and so forth. The current urbanization challenges facing China can be summarized as a problem of low quality urbanization. Urbanization in China epitomizes "incomplete urbanization" and is characterized by high input, consumption, emission and expansion. This is in contrast to desirable traits of "high quality" urbanization, characterized by low levels of input, quality, harmony, inclusiveness and sustainability. The proper resolution of these problems in the next round of development will directly determine the future of urbanization in China.

1.4.1 The Hukou Urbanization of Farmers: Slow and Difficult

As mentioned above, urbanization in China is a typical example of incomplete urbanization. Such incompleteness is mainly manifested by the fact that large numbers of rural migrant workers, suburban farmers who have received urban *hukou*, and farmers who live in areas recently transitioned to urban areas have yet to be fully integrated into urban society despite permanently living in towns and cities and counted as urban residents. These people remain outside the urban system and find themselves stuck in between urban and rural life -still with farmers' habits and characteristics, different than both urban and rural residents. This suggests a low rate of hukou urbanization (i.e., the process of making new registered city residents) (Chan 2009). With respect to the hukou urbanization of non-local rural migrant workers, there were 269 million rural migrant workers, including 166 million non-local ones, across China in 2013. These non-local rural migrant workers mostly work and live in urban areas, and although they are counted as urban residents, they face difficulties in accessing the same benefits and services available to urban residents, including employment, wages and benefits, childhood education, social security and affordable housing (Zhu 2007; Wong et al. 2007). With respect to the hukou urbanization of local farmers, the rate is also very low among suburban farmers who have received an urban *hukou* in the past few years, and farmers who live in areas recently developed into urban areas. An artificially high urban population has resulted from changes in China's system of administrative district (e.g., a large number of townships have been merged into towns and counties that have been upgraded to become districts) and the expansion of town- and city-administered areas, in particular. These areas are short of industrial support since their industrial structures have yet to transform. In reality, they remain rural areas since they lack the infrastructure required for towns and cities (Lu 2007).

In light of the aforementioned two factors, people with rural *hukou* account for a large part of the current urban population in China, and permanent urban residents with non-local *hukou* hold a large share of that population. In 2011, the urbanization rate of permanent residents registered on the household registration systems in their places of permanent residence was only 35%, which is 16.3 percentage points lower than the urbanization rate of all permanent residents. In 2010, the difference was only 10.5 percentage points. Data from the 2010 population census show that people with rural *hukou* accounted for 46.5% of China's urban population. Specifically, such people accounted for 36.1% of the total population of Chinese cities and 62.3% of that of Chinese towns. Today, urban residents with rural *hukou* have become dominant in China's urbanization process. Between 1978 and 2010, the net increase in China's urban population was 493 million people, including 262 million people with rural *hukou*, or 53.0% of the net increase. Meanwhile, China's urbanization rate increased by 31.76 percentage points, of which urban residents with rural *hukou* contributed 18.12 percentage points (see Table 1.6). In other words, China's urbanization rate would increase by only 13.64 percentage points—only 0.43 percentage points per year—if people with rural *hukou* were excluded from the calculations. This suggests that the hukou urbanization of rural migrants into Chinese towns and cities is much slower than ideal, resulting in a new dual structure in Chinese cities. This has led to several challenges. First, it has reduced the contributory role of urbanization in driving domestic demand and has had detrimental effects on improving the industrial structure and workers' competence. Second, it has created inequality between rural migrants and native urban residents in terms of rights and benefits, thus severely weakening social harmony and stability. Third, it has worsened the situation in which one's place of permanent residence differs from the place where one's *hukou* is registered, making it even more difficult to manage people. In 2010, there were

Table 1.6 Changes in China's urbanization rate, 1978–2010

	Total population (in million people)	Total urban population		Urban residents with rural *Hukou*	
		Size (in million people)	Share of total population (%)	Number (in million people)	Share of total population (%)
1978	962.59	172.45	17.92	48.01	4.99
2010	1339.72	665.70	49.68	309.59	23.11
Difference	377.13	493.25	31.76 percentage points	261.58	18.12 percentage points

Source: *Demographic Statistics Collection of the People's Republic of China* (1949–1985), the *Tabulation on the 2010 Population Census of the People's Republic of China* and *China Statistics Yearbook* 2011

already 226 million urban residents, or 33.7% of China's urban population, who lived in places other than where their *hukou* was registered.

1.4.2 Resource Intensity Versus Utilization Inefficiency

China has prioritized qualitative growth over qualitative improvement over the past few years in its urbanization drive. Specifically, China's fast urbanization has been based upon high consumption of energy and resources, at the costly expense of resource overuse and environmental degradation, resulting in inefficiency (Fang et al. 2013). The consumption of energy and resources has been increasing sharply. Figure 1.6 presents the consumption of total energy and the production of steel and cement during the urbanization process. It can be concluded that the increase of urbanization level was relatively consistent with resources and energy consumption in 1978–2000; while in 2000–2015, the consumption of energy and resources was significantly higher than the growth rate of urbanization level. The data indicates that from 1978 to 2000, China's urbanization level grew at an average annual rate of 3.25%, while the national average consumption growth rate of steel, cement and total energy was 6.56, 10.59 and 4.39% respectively; from 2000 to 2015, the annual growth rate of urbanization level was 2.96%, while the annual growth rate of consumption of steel, cement and total energy reached up to 13.56, 9.59 and 7.42% respectively. Furthermore, the total amount of resources consumption was even more astonishing. One percentage point increase in China's urbanization level in 1978–2000 has consumed steel with 86.23 million tons, cement with 340.13 million tons, and total energy with 1195.22 million tons of standard coal, respectively, while in 2000–2015,

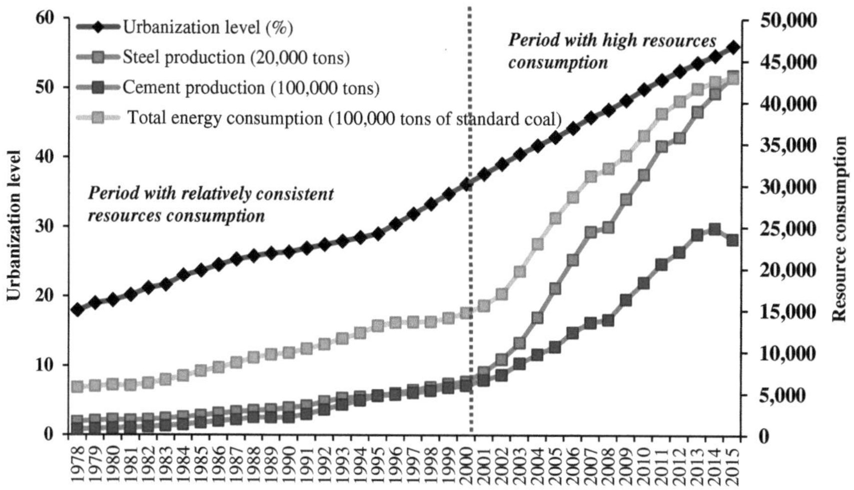

Fig. 1.6 Consumption of steel, cement and energy during the urbanization process. *Note* The data comes from China Statistical Yearbook, China Cement Association and China Iron and Steel Association

these figures has sharply increased to 386.26, 1170.43, and 2359.65 million tons of standard coal.

The supply of water resources is also far from meeting demand. Chinese cities saw water consumption increase sharply from 7.87 billion cubic meters in 1978 to 50.79 billion cubic meters in 2010, or an annual average increase in water consumption of 1.34 billion cubic meters. As a result, about 400 Chinese cities today suffer water shortages, including over 200 cities severely short of water. Most of these water-hungry cities have been excessively exploiting groundwater, resulting in faster land subsidence. Today, there are already over 50 Chinese cities suffering disastrous land subsidence. Many cities in the northern, northwestern and eastern regions face dropping groundwater levels, with visible appearance of large-area groundwater funnels and land subsidence (The General Report Authoring Team 2012).

Lastly, China's resources have been utilized inefficiently. China has made sufficient progress with respect to the efficiency of resource utilization, with energy consumption per 1000 yuan GDP decreasing from 1.568 tons of standard coal in 1978 to 0.0793 tons of standard coal in 2011. However, when compared with other countries in terms of energy consumption per $1,000 USD GDP, China is 2.3 times higher than the global average, 4.1 times higher than OECD member countries, 3.4 times higher than the United States, 5.7 times higher than Japan, 7.4 times higher than the United Kingdom and also higher than non-OECD countries. Given the current situation, it is impossible for China to comprehensively change the inefficient pattern of economic development and solve its perennial structural problems in a short period of time. Heavy pressures caused by the shortage of energy sources and water resources will continue to be an obstacle to China's further urbanization and economic development.

1.4.3 Growing Pollution and Environmental Degradation

China's path to urbanization has long been characterized by high consumption and high emission rates. Today, China is facing mounting challenges in terms of resources and the environment. There are three aspects to these growing challenges. First, there has been a fast increase in pollution. Between 2001 and 2010, the total amount of solid, liquid and gaseous waste from industry grew by 11.4, 14.2 and 4.0% per year respectively. In the foreseeable future, the total amount of primary pollutants will remain high. China is already close to its highest carrying capacity in terms of the environment. Second, the environment is under heavy pressure. Globally, there has been a stronger call for carbon emissions reduction as climate change has become a growing concern. Accordingly, all countries are under increasing pressure to reduce their carbon emissions. The Chinese government, for example, has promised to reduce carbon emission levels by 40–50% by the end of 2020. Furthermore, China's 12th Five-Year Plan includes compulsory targets with respect to farmland area retention, resource and energy consumption as well as reduction of main pollutants. Greater efforts by China to conserve energy and reduce emissions will bring an end to the traditional, inefficient pattern of economic development characterized by high consumption, high emission rates and fast expansion in pursuit of fast growth. Third, Chinese cities suffer from severe environmental problems, such as air and water pollution and municipal solid waste (MSW). Two-thirds of Chinese cities are non-compliant in terms of air quality, according to the *Ambient Air Quality Standard* revised in February 2012. In 2011, non-compliant cities in terms of ambient air quality still comprise 11% of the 325 Chinese cities at the prefecture and higher levels, including the seats of governments of some prefectures, autonomous prefectures and leagues, as well as municipalities directly under provincial governments. 227 counties and cities, or 48.5% of all the 468 counties and cities monitored, have experienced acid rain. Sewage discharge increased by 37.4% between 1991 and 2011 in Chinese cities, and by 84.1% in county seats between 2001 and 2011. Today, nearly 20% of sewage from Chinese cities, 30% of sewage from county seats and the vast majority of sewage from towns receive no effective waste treatment before being discharged directly into rivers, lakes or seas. This has severely contaminated water resources and thus threatened urban drinking water safety. In 2011, 55% of all the 4727 groundwater monitoring points in 200 Chinese cities reported poor or very poor water quality. In the 113 Chinese cities identified as major cities in China's environmental protection program, 9.4% of the total amount of drinking water was reported as non-compliant with safety quality standards. More shocking are reports of severe water safety problems such as hypertrophication in the Taihu and Dianchi Lakes and heavy metal pollution in the Xiangjiang River. Additionally, most Chinese cities are now beleaguered by a growing amount of municipal solid waste (MSW).

Table 1.7 Annual average growth: urban population versus land used for urban development in China

Period	Urban population (%)	Built-up area (%)	Land use for urban development (%)
2001–2005	4.13	7.70	7.99
2006–2012	3.43	4.94	5.02
2001–2012	3.72	6.08	6.25

Note Shanghai data on land area used for urban development in 2005 is unavailable, thus the average of its 2004 and 2006 numbers is used instead
Source: *China Urban-Rural Construction Statistical Yearbook* (2012) and *China Statistics Yearbook* (2013)

Across China, there is over 7.0 billion tons of MSW covering over 53,300 ha of land. Moreover, the amount of MSW is growing by 4.8% per year on average. Two-thirds of medium-sized and large Chinese cities have overwhelming amounts of MSW.

1.4.4 Town and City Sprawl

Chinese towns and cities have experienced unprecedented and even uncontrolleable levels of urban sprawl amid recent industrialization and urbanization frenzy (Lu 2007). This is in part due to the fact that land urbanization has progressed much faster than population urbanization. Between 2001 and 2012, the total amount of build-up and total amount of land allocated for urban development in China increased by 6.08 and 6.25% per year on average respectively, whereas the urban population grew by only 3.72% per year on average. In the 10th Five-Year Plan period, in particular, the aforementioned metrics increased by 7.70 and 7.99% per year on average, much higher than the 4.13% annual average growth rate of the urban population (see Table 1.7). This suggests that land consumption in Chinese cities has far outpaced population growth in recent years. In other words, land urbanization has progressed at a much faster rate than population urbanization in China.

Town and city sprawl is also a phenomenon across China. The total size of built-up areas increased by 5417.7 km^2 between 1981 and 1990, by 9583.6 km^2 between 1991 and 2000, and by 23,126.5 km^2 between 2001 and 2012. From 2001 to 2012, in particular, built-up areas increased by 1927.2 km^2 per year on average, which was 3.58 and 2.01 times higher than in the first two periods respectively. In terms of average growth per city, the average size of built-up areas per city grew from 30.4 km^2 in 1996 to 69.4 km^2 in 2012. In the meantime, the average land area used for urban development per city increased from 28.5 to 69.6 km^2. This represents a growth rate of 128.5 and 144.1% respectively (see Fig. 1.7). Moreover, urban population density

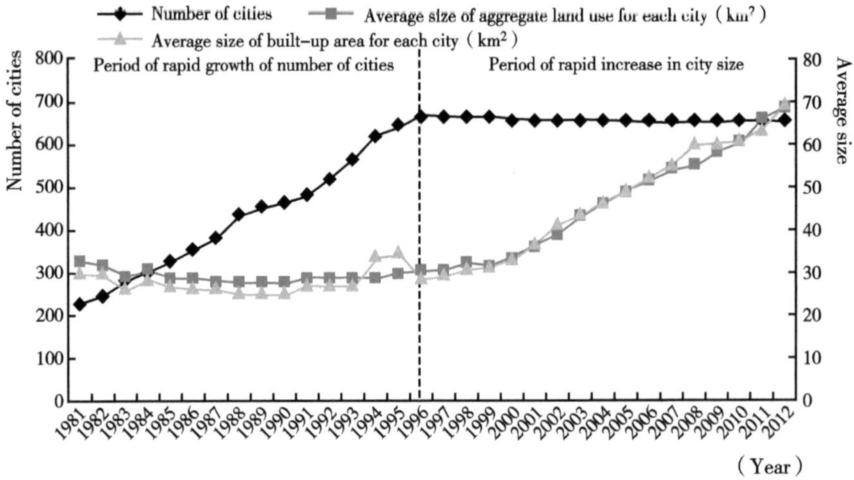

Fig. 1.7 Growth of the number and size of cities in China. *Note* Regarding the land area used for urban development in 2005, Beijing and Shanghai data is unavailable and the averages of their 2004 and 2006 numbers are used instead Source: *China Urban-Rural Construction Statistical Yearbook* (2008–2012)

in built-up areas[1] decreased from 8500 people/km^2 in 2000 to 7300 people/km^2 in 2011. The average size of built-up areas per capita increased from 117.1 m^2/person in 2000 to 137.3 m^2/person in 2011, a figure much higher than the 80–120 m^2/person standard specified in the *Urban and Rural Planning Law*, and one that reached or even surpassed that of developed countries (Lu and Yao 2007). To a certain extent, the fast growth of China's urban economy in recent years has been mainly driven by urban area expansion.

Lastly, some large Chinese cities have been expansion is upgrading counties and/or county-level cities to districts, as wells as building new districts and/or industrial parks. This somewhat unreasonable urban area expansion is partly attributable to insufficient restrictiveness of urban development planning, which has been exploited by self-interested urban planners. New urban districts tend to be planned with several hundred or even over a thousand square kilometers in area, such as the Shanghai Pudong New Area (1210.41 km^2), Tianjin Binhai New Area (2270 km^2) and Chongqing Liangjiang New Area (1200 km^2). In some Chinese cities, industrial parks may be planned with over 100 km^2 in area. In reality, fast urban sprawl in China has led to a sustained decrease in farmland area and quality. In 2008, for example, the total area of farmland used for urban development across China was as high as 191,600 hectares, or 68.9% of the total decrease in farmland area that year (Wei and Zhang 2011). Most of the used land was high-quality farmland. Although China has implemented a policy aimed at replacing used farmland with reclaimed

[1]The population refers to China's urban population, and the size of built-up areas is the total area of Chinese towns, county seats and cities.

land of ostensibly the same area and quality, such farmland is actually often made up for by land with lower quality. With the rapid increase in the size of built-up areas and in the number of various industrial parks, the area of high-quality farmland in plains areas or near human settlements has been decreasing rapidly. If China does not undertake efforts to combat this phenomenon, the total area of farmland in China will drop below the 120 million-hectare red line. This will seriously threaten China's food security.

1.4.5 The Urban System Lacks Reason

Due to the fact that resources are allocated according to a region's administrative level, a disproportionate amount of China's resources and development opportunities are unfairly concentrated in major administrative centers such as Beijing, other municipalities directly under the Central Government, and cities at the sub-provincial level and provincial capitals. This has resulted in polarization of China's urban system. First, large cities have been expanding rapidly, resulting in fast local population growth and the clustering of industrial enterprises amid rapid urban area expansion. Accordingly, these cities now face severe problems such as energy and resource shortages, urban sprawl, traffic jams, air pollution, continuously declining living standards, and growing social conflicts (Zhao et al. 2003). Second, the percentage of small and medium-sized cities in China, as well as the percentage of China's total population living in small and medium-sized cities, have been decreasing. In comparison, large cities have been increasing in number amid a general trend of county-level cities being upgraded into districts and entirely new districts being established. Specifically, between 2002 and 2012, the percentage of large cities with a population of at least 500,000 people (as a total of all Chinese cities) has been growing, as has the percentage of China's total population living in large cities. In contrast, the percentage of medium-sized cities as a total of all Chinese cities has been dropping, as has the percentage of small cities in both indicators. Third, there are a large number of towns across China, but they tend to be small and financially weak. The number of Chinese towns increased from 2173 in 1978 to 20,117 in 2013, an average increase of over 500 towns per year. Unfortunately, towns usually suffer from the siphoning effect of large cities, and thus are deprived of many development opportunities. In addition, towns themselves lack industrial support and lag behind in terms of infrastructure and public services. As a result, towns have insufficient ability to rapidly develop on their own. In 2012, 175.33 million people lived in the built-up areas of towns, which comprised 37,145 km^2 in total land area. In other words, there were only 10,200 people living in the built-up areas of each town on average, and such areas were merely 2.16 km^2 per town on average. Furthermore, Chinese cities face severe problems such as blind imitation of each other, dominance by low-end industries, and uniform industrial structures. As a short summary, problems with Chinese towns and cities mainly consist of unclear positioning, functional overlap and unclear division of labor. Additionally, the relationship between central cities and the surrounding small and medium-sized cities is more often than not characterized by the central city depriving the smaller cities of resources, leading to a polarization, rather than ripple, effect. In the future, it is important that China leverage the advantages of cities of different sizes so as to gradually create an urban system characterized by clear division of labor, complementary functionality, collaboration and enhanced competitiveness.

1.4.6 Unbalanced Regional Development

China's unbalanced regional development is manifested in three aspects. The first is the obvious interregional difference in urbanization rate. As a result of natural and historical factors, the urbanization rate across China clearly decreases from the east to the west. In 2012, for example, Shanghai, Beijing and Tianjin were the top three places with the highest urbanization rate among all of China's municipalities and provinces, and the top eight were all located in eastern and northeastern China. Moreover, the highest urbanization rate (Shanghai) was 3.93 times higher than the lowest (Tibet). The average urbanization rate was 69.06% among the top ten municipalities and provinces and 40.38% among the bottom ten, an astonishing gap of 29 percentage points. The second aspect is China's unbalanced spatial distribution. As a result of natural conditions and regional development policies, the vast majority of Chinese cities are in southeastern China, according to the Aihui-Tengchong Line, a concept developed by Chinese demographer Hu Huanyong to illustrate the spatial distribution of the Chinese population. In 2012, China's western regions, which account for 72% of China's land area, had only 25.2% (165 cities) of all Chinese cities, and accounted for a mere 19.5% of China's urban population. In contrast, the eastern regions, which account for only 10% of China's land area, had 35.6% (233 cities) of all Chinese cities, and accounted for 48.8% of China's urban population (see Table 1.8). With respect to population per city, there are 26 Chinese cities with a non-agricultural population of more than two million people, 14of which (53.8%) are in the eastern regions. On the other hand, there are 495 Chinese cities with a non-agricultural population of fewer than 500,000 people, 270 of which (54.5%) are in China's central and western regions. The significantly unbalanced regional placement and population distributions of Chinese cities will inevitably lead to heavy pressure on resources and the environment in some areas, interregional movement of energy sources and commodities which will add to socioeconomic development costs, and an interregional imbalance in terms of economic development. The third aspect is the wide interregional gap in the level of urban development. In 2010, for example, western cities were only 30.4, 68.3, 44.3, 28.9, 45.2 and 47.4% as high as eastern ones in terms of per-capita industrial added value, investment in fixed assets, year-end balance of household savings, local revenue—general budgeted income and expenditure, as well as consumer goods retail sales, respectively. Between 2005 and 2010, the east-west gap in disposable income per urban resident widened from

Table 1.8 Chinese city population mix by region, 2012

Item		Total	>2.0 million	1.0–2.0 million	0.5–1.0 million	0.2–0.5 million	<0.2 million
China	City count	655	26	39	95	250	245
	Population share (%)	100.0	33.1	15.8	18.8	22.8	9.5
East	City count	233	14	22	38	104	55
	Population share (%)	48.8	20.2	8.9	7.7	9.9	2.1
Northeast	City count	89	4	5	14	26	40
	Population share (%)	12.4	3.8	1.7	2.7	2.4	1.7
Central	City count	168	4	6	27	61	70
	Population share (%)	19.3	3.6	2.5	5.2	5.4	2.7
West	City count	165	4	6	16	59	80
	Population share (%)	19.5	5.6	2.6	3.2	5.1	2.9

Note City population sizes here in are counted and grouped by non-agricultural population
Source: *China Population & Employment Statistics Yearbook* (2013)

4592 yuan to 7467 yuan. For example, the disposable income per resident in Shanghai was 2.41 times higher than in Gansu Province.

Unbalanced urban and rural development is primarily manifested by an urban-rural income gap that stays high. Despite a narrowing urban-rural income gap across China in recent years, the ratio of disposable income per urban resident to net income per rural resident in China was still as high as 3.10 to 1 in 2012 (see Fig. 1.8). It should be noted that the net income of farmers includes in-kind income, which is not insignificant. Yet part of their cash income must be spent on purchasing agricultural equipment, chemical fertilizers, pesticides, seeds and other production materials. If both of these factors are deducted from farmers' net income, then actual disposable income per capita is still 4–5 times higher for urban residents than rural residents. Globally, a Gini coefficient of 0.4 is considered the warning line. A Gini coefficient of over 0.4 means that wealth is excessively concentrated and the gap between the rich and poor is unacceptably wide so as to be very likely to lead to social instability. Since the 1990s, China's Gini coefficient has been very high. NBS data shows that the Gini coefficient for Chinese residents remained between 0.47 and 0.49 between 2003 and 2013. After reaching a peak value of 0.491 in 2008, the coefficient gradually decreased but was still as high as 0.474 in 2012. A Gini coefficient this high signals great danger. A persistently wide urban-rural gap in income has become a major obstacle to China's urbanization and socioeconomic development.

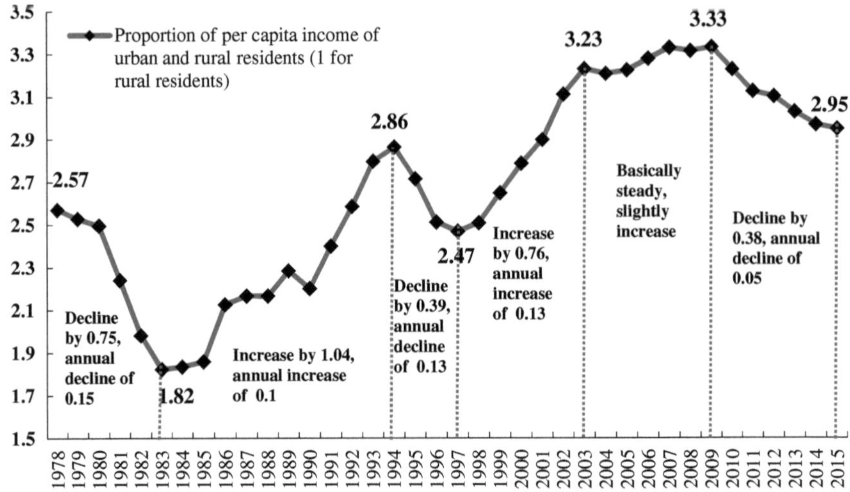

Fig. 1.8 Changes in income gap between urban and rural areas, 1978–2015 Source: *China Statistics Yearbook* (in previous years)

1.4.7 Non-inclusiveness in Urbanization

The non-inclusiveness of China's urban development is also becoming an increasingly severe problem in the urbanization process (Zhuang 2008). This problem is manifested in income growth that lags behind economic growth, a widening gap among social classes, growing poverty in urban areas, and region-specific deprivation, among other issues.

First, income growth lags behind economic growth. China's fast economic growth has long relied heavily on investment, and growth in income and consumption has lagged far behind increase in government revenue and GDP. In other words, resident income and consumption growth is much slower than GDP growth, which in turn is much slower than investment and revenue growth. Between 2001 and 2010, China achieved a 20% average annual increase in revenue and a 10.5% annual average increase in GDP based on a 23% annual average increase in fixed asset investment. This is in sharp contrast to the annual average increase of 9.7% in average disposable income per urban resident, 8.0% increase in consumption per resident, and 7.0% increase in net income per farmer. In this period, both fixed asset investment and revenue grew nearly twice as fast as GDP, which in turn grew at a rate 0.8 percentage points higher than that of income per urban resident and 2.5 percentage points higher than that of consumption per resident.

Second, the gap among social classes is widening. This is primarily manifested by a widening gap between urban and rural residents. In 2010, 86.8% of fixed asset investment across China was invested in urban areas, despite the fact that less than half of China's total population lived in urban areas, even accounting for rural migrant workers as part of the urban population. Consequently, the average income per rural resident grew at a rate 2.7 percentage points lower than that of the average income per urban resident between 2001 and 2010. This demonstrates that Chinese farmers do not yet adequately share the benefits of China's fast economic growth. Furthermore, issues in rural areas-including land loss and an imbalanced rural population consisting predominately of the elderly and the left-behind wives and children of rural migrant workers-constitute major concerns with respect to social instability. Second, rural migrant workers have been treated as second-class citizens in Chinese cities despite their many contributions to urban development. These workers continue to find it difficult to fully integrate into cities as a result of discrimination in employment, income, education, medical and health care, and so forth. The "one city, two systems" policy has led to second-class treatment and marginalization of rural migrant workers (Lu 2000). Third, the rise of the "new dual structure" in cities has created the growing phenomenon of residential space differentiation between rich and poor neighborhoods. In the so-called "rich neighborhoods," members of the high-income class—who are the minority of the total population—live in upscale apartment complexes or houses. A growing number of poor neighborhoods, "villages in the city" and shanty towns, have appeared in old urban areas and suburbs. There is a fear that the increasing residential space differentiation and resulting spatial segregation will trigger a series of social conflicts and thus harm China's goal of building a harmonious society.

Third, there is growing poverty in urban areas. Today, poor people in Chinese towns and cities are mostly low-income employees, the retired, the unemployed, people with no income or financial support and no ability to work for a living, and non-local rural migrant workers (Liang and Fang 2011). In 2011, the average disposable income per capita of the 10% lowest-income households was only 31.5% as high as the national average; that of the 5% neediest households was only 24.8% as high as the national average. On the flip side, the top 10% of household income earners had incomes that were 8.6 times higher than the lowest 10%, and 10.9 times higher than the lowest 5% of income earners, in terms of average disposable income per capita according to a sample survey by the NBS among 65,655 urban households across China. Growing poverty in urban areas has become a major challenge for healthy urbanization and socioeconomic sustainability in China (Tang 2002).

Lastly, region-specific deprivation is a growing concern. A domino effect of economic deprivation has occurred due to unfair allocation of opportunities and benefits amid fast urbanization. Enabled by policy loopholes and administrative coercion, some advantaged groups and regions have been able to seize resources, opportunities and benefits, while disadvantaged groups and regions are left to bear the costs of resource consumption and environmental pollution. There are several examples of this deprivation. Large cities monopolize resources and opportunities at the expense of small and medium-sized cities. Farmland and farmers' livelihoods are being lost to the development of holiday resorts and university zones. Rural and urban residents lose their properties and living spaces as a result of real estate development and "village in the city" reconstruction projects. Developed regions deprive less developed regions of resources and development opportunities. Regions with a large influx of rural migrant workers deprive the regions from which the rural migrant came from resources and labor force. Region-specific deprivation is resulting in uncoordinated spatial development, unfair resource allocation, partial or even total policy failure, and a widening income gap between the rich and poor. This is ultimately impeding the Chinese goal of building a harmonious society (Fang and Liu 2007).

1.4.8 Homogenization in Urban Design and Management

Throughout the process of urbanization in China, urban development nationwide has adopted a near-homogeneous style, leading to severe loss of diversity. This is a result of limited financial input, insufficient knowledge of local features and cultures, and short-term thinking. There are two factors in play behind homogenization in urban development across China. First is the phenomenon of large-scale demolition and reconstruction across a large number of cities, with woefully inadequate protection of locally unique, historic and landmark buildings and disregard for culturally distinctive villages and towns. Urban reconstruction projects typically focus on speed and trying to look as modern as possible while neglecting the inheritance and innovation of traditional cultures. Consequently, numerous old buildings, streets and blocks with cultural and historical significance have disappeared completely. In addition, new villages tend to be planned and constructed to copy urban functions. As a result of the copying mentality, locally distinctive old memorial temples, buildings, gardens and villages are damaged or even destroyed. The second factor is the current trend of copying foreign styles in the design of buildings and apartment complexes, rather than integrating local features. Many Chinese cities grossly misunderstand modernity and internationalization and blindly imitate foreign architecture and overemphasize unique, peculiar forms of building design in the pursuit of quick success. This is why new urban areas and buildings in China usually look almost identical. Compounding on the impulse to emulate foreign styles, local governments usually prioritize meeting economic targets over anything else, resulting in white-elephant projects that feature excessive land use and demonstrate ineffective urban planning and management.

1.5 Review of China's Urbanization Strategy

Urbanization is a necessary approach for a country transitioning from an agrarian to industrial economy, a general trend of societal development, and an important mark of a modern country. Despite several remarkable achievements, urbanization in China faces numerous challenges. It is therefore important, both theoretically and practically, to review China's previous urbanization strategy, identify lessons learned, and seek to rapidly improve China's urbanization strategy so as to systematically promote urbanization. There are several lessons to be learned from China's past experience.

First, urbanization in China has not been fully people-centric. Urbanization is essentially a process where a traditional agrarian society evolves into a modern, civilized, industrial and information-based society. Urbanization thus requires a thorough change in the means of production and social lifestyles. However, this is lacking in China. The current income and consumption levels of rural and urban residents in China lag behind the level of economic growth. There is a widening income gap between rural and urban residents, as farmers continue to lack urbanization-relevant benefits. The hukou urbanization of rural migrants is progressing at too slow a pace, resulting in inequality between them and native urban residents with respect to rights and benefits. In other words, urbanization in China has failed to prioritize the ability of all urban residents to equally enjoy modern urban civilization and the benefits of urban life. This is a typical example of incomplete urbanization. If people-centric issues are not solved quickly, urbanization will slow, and may even trigger social unrest. To achieve complete urbanization, it is critical to increase the income of urban and rural residents, narrow the urban-rural gap, create diversified cost sharing mechanisms, and speed up the hukou urbanization of farmers. This will allow urban and rural residents to equally share the benefits of urbanization.

Second, urbanization in China has been a typical example of inadequate control. Rapid urbanization has been based upon excessive resource and energy consumption, land use and waste generation, at the expense of natural resources and the environment. The overall benefits have been low. Such an ineffective pattern is unsustainable. In its original sense, urbanization is also a process of clustering industries and factors of production to maximize resource efficiency. However, urbanization in China today is characterized by high consumption/emissions intensity and rapid expansion, as described by the aforementioned ineffective pattern. It is critical that China act quickly to change its resource-intense urbanization approach to a more sustainable urbanization strategy characterized by low resource intensity, environment friendliness and improved efficiency.

Third, China has improperly handled various relations in the process of urbanization, prioritizing rate over quality, urban development over management, production over enjoyment of life, economic growth over social justice and development over cultural protection. Going forward, China should better balance the priorities of rate and quality, urban development and management, production and life, economy and society, as well as development and protection. This is critical for exploring a new, China-Model approach to urbanization.

Fourth, urbanization in China lacks strong support from urbanization-relevant systems and mechanisms. The existing household registration system, along with current land, employment, social security, tax/finance and administrative division management systems, have created an urban-rural split, making it difficult for urban and rural residents to equally access basic public services. Such systems have also hindered the hukou urbanization of rural migrants and the integration of urban and rural development, while resulting in unfair interregional resource allocation, a widening interregional gap and unbalanced urban distribution. Institutional barriers are the primary cause of the deep problems with urbanization in China and thus make healthy urban development impossible. It is therefore necessary to center relevant efforts on the major tasks and objectives for China-Model urbanization and to remove the aforementioned hindrances by accelerating the comprehensive urban and rural reforms and establishing or improving the household registration, land, employment, social security, tax/finance and administrative division management systems. To comprehensively improve the quality of urbanization, China should take measures to facilitate the free flow and equitable exchange of the factors of production between urban and rural areas as well as the optimal allocation of public resources. It should also focus on advancing population and land urbanization and hukou urbanization.

In brief, China must change its approach to urbanization as early as possible by replacing the traditional method of urbanization with a new, China-Model approach characterized by sustainable development, high effectiveness and efficiency, low carbon emissions, inclusiveness/sharing and integrated urban and rural development. Great efforts should be made to improve the quality of urbanization, reduce the costs of resources and the environment in the process of urbanization, and lay a solid basis for building a socialist society with Chinese characteristics.

References

Cai, F., & Wang, D. W. (1999). China's economic growth: sustainability and labor contribution. *Economic Research Journal, 10,* 62–68.

Chan, K. W. (2009). The Chinese *hukou* system at 50. *Eurasian Geography and Economics, 50*(2), 197–221.

China Development Research Foundation. (2010). *China development report: China's new urbanization strategy for human development*. Beijing: The People's Press.

Fang, C. L. (2009). Three decades of China's urbanization and urban development since it began reform and opening up. *Economic Geography, 29*(1), 19–25.

Fang, C. L., & Liu, H. Y. (2007). Region-specific deprivation and regulatory path in rapid urbanization. *Acta Geographica Sinica, 62*(8), 860–948.

Fang, C. L., Guan, X. L., Lu, S. S., Zhou, M., & Deng, Y. (2013). Input-output efficiency of urban agglomerations in China: An application of data envelopment analysis (DEA). *Urban Studies, 50*(13), 2766–2790.

Fang, C. L., Yao, S. M., Liu, S. H., et al. (2011). *The report on urban agglomerations in China 2010*. Beijing: Science Press.

Liang, D. (2012). Urbanization as a significant factor for increasing domestic demand. *Shanghai Securities News*. March 26, 2012.

Liang, H. M., & Fang, C. L. (2011). Poverty in Chinese cities: Basic characteristics and the way out of poverty. *Human Geography, 6,* 61–66

Liu, G. S. (2005). The concepts of city, town and city, countryside and village: Understanding and usage. *Journal of Shandong University of Technology (Social Science Edition), 21*(1), 54–57.

Liu, Y. (2011). Urbanization in China: History, problems and trends. *Research on Economics and Management, 3,* 20–26.

Lu, X. Y. (2000). How to go beyond policies that cause urban-rural split. *Dushu, 5,* 3–9.

Lu, D. D. (2007). Urbanization in China: Process and spatial expansion. *Urban Planning Forum, 4,* 47–52.

Lu, D. D., & Yao, S. M. (2007). Urbanization in China from a scientific perspective. *Human Geography, 4,* 1–6.

Qiu, Y. (2013). Urbanization's effect on technological innovation—Empirical research based on 1990–2010 panel data about Chinese regions. *Population Science of China, 33*(1), 35–36.

Shi, H. N. (2012). Urbanization's effect on human capital accumulation in China. *China Soft Science, 3,* 117–127.
Tang, J. (2002). Current poverty in Chinese cities: reasons and status. *China Cadres Tribune, 3,* 22–25.
The General Report Authoring Team. (2012). In J. H. Pan & H. K. Wei (Eds.), *Annual report on urban development of China* (Vol. 5). Beijing: Social Sciences Academic Press (China).
Wang, M. T. (2013). The sequence of administrative levels of cities and the urban & rural planning system. In *Collaborative planning in the era of cities—the collection of papers at the Annual National Planning Conference 2013.*
Wang, X. L., & Xia, X. L. (2000). Urbanization's role in economic growth. In X. L. Wang & G. Fan (Eds.), *Sustainability of China's economic growth*. Beijing: Economic Science Press.
Wei, H. K. (2005). How to understand "healthy urbanization is an important part of restructuring". *People's Daily*, January 19, 2005.
Wei, H. K. (2014). *A new road to urbanization with Chinese characteristics*. Beijing: Social Sciences Academic Press (China).
Wei, H. K., & Zhang, Y. (2011). Toward green urbanization in China: Ideas and measures. *Economic Review, 9,* 15–19.
Wong, D. F. K., Li, C. Y., & Song, H. X. (2007). Rural migrant workers in urban China: living a marginalized life. *International Journal of Social Welfare, 16*(1), 32–40.
World Bank Group. (1997). *China 2020: Development challenges in the new century*. Beijing: China Financial & Economic Publishing House.
Yao, S. M., Chen, Z. G., & Zhu, Y. M., et al. (2006). *The urban agglomerations of China*. Hefei: Press of University of Science and Technology of China.
Zhao, S. X. B., Chan, R. C. K., & Sit, K. T. O. (2003). Globalization and the dominance of large cities in contemporary China. *Cities, 20*(4), 265–278.
Zhou, Y. X. (1989). The concept of town and city and statistical metrics of the urban population in China. *Population & Economics, 1,* 9–13.
Zhou, Y. X. (2007). *Urban geography*. Beijing: The Commercial Press.
Zhou, Y., & Sun, X. Z. (2012). The road to urbanization in China: Review and countermeasures. *China Population, Resources and Environment, 22*(4), 56–59.
Zhu, Y. (2007). China's floating population and their settlement intention in the cities: Beyond the *hukou* reform. *Habitat International, 31*(1), 65–76.
Zhuang, J. (2008). Inclusive growth toward a harmonious society in the People's Republic of China: Policy implications. *Asian Development Review, 25*(1), 22–33.

Chapter 2
The Essentials of the New Chinese Approach to Urbanization

In recent years, academic explorations of China's approach to urbanization have shifted in focus from "urbanization with Chinese characteristics" to "new urbanization," to "new urbanization with Chinese characteristics." The notions of urbanization with Chinese characteristics and new urbanization are not mutually exclusive but an organic whole. The "New China-Model" approach to urbanization organically combines these two preceding concepts. This new approach is characterized by consideration for Chinese conditions and characteristics at different development stages, as well as a people-first mentality, high efficiency, green development and low carbon emissions, integrated urban and rural development, strong coordination of industrialization/urbanization/agricultural modernization with information technology (IT) applications, and cultural inheritance.

2.1 Scientific Foundation of the New Chinese Approach to Urbanization

In recent years, Chinese scholars have engaged in in-depth discussions on China's path to urbanization, including "urbanization with Chinese characteristics" and "new urbanization." The term "urbanization with Chinese characteristics" was already coined by the mid-1980s, and continued as an approach into the mid-2000s. In November 2002, a report to the 16th Party Congress stated, "It is essential to … persist in the coordinated development of large, medium and small cities and small towns along the path to urbanization with Chinese characteristics." The report to the 17th Party Congress in October 2007 included references to "a path of urbanization with Chinese characteristics," "the path of independent innovation with Chinese characteristics," "the new path of industrialization with Chinese characteristics," "a path of agricultural modernization with Chinese characteristics," and "the path of political development under socialism with Chinese characteristics." These formed

H. Wei, *Urbanization in China*, Research Series on the Chinese Dream and China's Development Path, https://doi.org/10.1007/978-981-13-1408-7_2

the basic elements of "the path of socialism with Chinese characteristics." This same report noted that "Taking a path of urbanization with Chinese characteristics will promote balanced development of large, medium and small cities and small towns based on the principles of balanced urban and rural development, rational resource distribution, land conservation, provision of a full range of functions, and assistance from larger cities to smaller ones."

Despite this lofty rhetoric, scholars have yet to reach consensus on what "urbanization with Chinese characteristics" really means. Jian (2003) believes that a path to urbanization with Chinese characteristics should be defined by an approach where urbanization, industrialization and modernization are carried out at a similar pace; where urbanization comes in a diversity of forms; where towns and cities of varying sizes develop in a coordinated manner; where government-guided and top-down initiatives are market-oriented and made in combination with bottom-up initiatives; and where towns and cities are developed logically with a focus on quality. Xiao et al. (2008) believe that a path to urbanization with Chinese characteristics refers to an approach that takes into account China's primary characteristic of a huge population with relatively insufficient farmland and is based on the objectives of balancing urban and rural development, ensuring rational distribution, conserving land resources, providing a wide range of functions, and getting larger cities to help smaller ones. They believe that this approach will lead to the rational distribution of resources, coordinated development of towns and cities, resource and environmental sustainability, rural modernization, and development diversification to suit regional characteristics. This in turn will create a new landscape of urban development characterized by resource conservation, eco-friendliness, economic efficiency and social harmony. At the kick-off ceremony for the fall semester of the Chinese Academy of Governance, and at a subsequent urbanization-themed workshop for mayors and provincial and ministerial officials, Chinese Vice Premier Ma (2012) noted that an approach to urbanization with Chinese characteristics should be based on a new model of urbanization that is informed by the Chinese government's Scientific Outlook on Development, characterized by efficient use of resources, diversity in development, high levels of coordination among industrialization, urbanization, and agricultural modernization, regulation by both the market and the government, and a commitment to putting real people and their real interests first.

There is a common thread to the various interpretations of urbanization with Chinese characteristics. It must comply with both the universal laws of urbanization and China's primary characteristics and cultural traditions, and must be customized to meet the country's needs at different stages of development and its primary characteristic as a large country (Niu 2010). To synthesize the views of various scholars, a path to urbanization with Chinese characteristics is one that satisfies at least the following five requirements (Zhou 2009; Niu 2010; Wu 2010; Gao et al. 2010).

First, urbanization should be based upon China's primary characteristic of relatively insufficient amount of resources per capita, and thereby should feature full use of land, high density of towns and cities and high efficiency. Second, urbanization should be conducted in a region-specific manner that fully reflects interregional differences in natural conditions, socioeconomic development and resource endowment. Third, urbanization should take place at a steady, reasonable rate in line with the limits of job and economic growth. China is a populous country, and every percentage point increase in the urbanization rate requires the creation of more than ten million jobs. The hukou urbanization of rural migrants should also be a priority. Fourth, urbanization should be based upon China's characteristics as a large country and one still in the early stage of socialism. Urbanization should proceed in a way that allows all stakeholders to play roles for which they fit, and that explores all possible patterns and paths for the coordinated development of towns and cities of varying sizes. Fifth, urbanization should be carried out in accordance with the core socialist principle that calls for the "elimination of serious inequality and the pursuit of shared prosperity." It is, therefore, imperative to pursue integrated urban and rural development so that the benefits brought by urbanization can be equally enjoyed by urban and rural residents alike.

The concept of "new urbanization" is not in fact entirely new. Former Premier Wen Jiabao called back in May 2007 for "a new type of approach to urbanization". And then in 2013 it was declared during the Central Economic Work Conference that the country needs "take new approach to urbanization that is effective, smart, green and low carbon." Scholars in China have also written about the topic. For Luo (2010) new urbanization has the following features: it puts people and their interests first, it is fueled by a new type of industrialization that facilitates the modernization of cities, agglomeration, green development, comprehensive planning, scientifically-sound development, high efficiency, function diversity, social harmony, retention of distinctive local styles, integrated and coordinated development for both urban and rural areas and for towns and cities of varying sizes. Zhang (2013) believes that the path to new urbanization has four defining features. First is the full and interactive coordination of industrialization, urbanization, IT application and agricultural modernization. Under this principle, industries and cities develop in an integrated manner; urban and rural areas develop together under the principle of comprehensive planning; and rural civilization is preserved. Second is the coordination of population and economic resources with environmental factors, in which efficient, smart, green and low-carbon development is promoted. Third is the compatibility of city plans and designs with regional economic and industrial development, in which development of towns and cities of varying sizes is coordinated. Fourth is an emphasis on human development that is comprehensive, inclusive and harmonious. Ni (2013) defines new urbanization as a path to sustainable, integrated urban and rural development that it is guided by the government's Scientific Outlook on Development. In other words, it is centered on population urbanization and powered by IT applications, agricultural modernization and a new type of industrialization, and is carried out in a market-oriented, quality-centric manner under government guidance. Qiu (2012) believes that new urbanization differs from traditional urbanization in six respects. First, instead of prioritizing the development of large cities, new urbanization seeks to achieve both urban and rural development that is complementary and coordinated. Second, it emphasizes low energy consumption over high energy consumption. Third, it shifts the focus from quantity to quality. Fourth, it emphasizes a low environmental impact. Fifth, it displays effective control rather than lack of control. Sixth, it departs from the adage of "let a few get rich first" and turns to the goal of more equitable distribution to build social harmony.

The rise of the concept of new urbanization can, therefore, be seen as a means to address the problems created by traditional urbanization. Systems such as centralized procurement and distribution of agricultural products, household registration that divides rigidly the urban and the rural areas, and people's commune were very much the products of the traditional development strategy under which, cities and heavy industry were given priority. As part of this strategy, the Chinese government conducted industrialization by extracting resources from agricultural and rural areas, while restricting the migration of rural residents into towns and cities. This strategy did lead to certain success, but it made coordination and interaction among industrialization, urbanization and agricultural modernization impossible and led to a bifurcated social structure that put a rigid division between the urban and the rural areas (Wang et al. 2011). In summary, new urbanization is associated with a new scientific outlook compared to traditional urbanization (Qiu 2010; Luo 2010). First, the focus is shifted away from increasing quantity/size toward improving quality. Second, the emphasis is on coordinated socioeconomic development rather than pure economic development. Third, the priority is on achieving integrated urban and rural development in a well-coordinated, complementary manner as opposed to just urban development.

As can be seen through this review, Chinese scholars in recent years have studied "urbanization with Chinese characteristics" and "new urbanization" in different manners, but all of them treat the two concepts as distinct and separate from each other. Urbanization with Chinese characteristics can readily be applied to both new and traditional urbanization as long as it is typical of China. By contrast, the patterns and practices associated with the idea of new urbanization may well suit countries other than China as well. The point is that the path to urbanization with Chinese characteristics and the path to new urbanization are not only not mutually exclusive but in fact mutually complementary. China must therefore organically combine them into an approach that is both new and distinctly Chinese (Wei 2010). There are two components to this approach to urbanization. On one hand, China must abandon the traditional urbanization mindset and embrace the new urbanization mindset guided by the criteria laid out in the Scientific Outlook on Development. On the other hand, China must continue to embrace urbanization with Chinese characteristics by taking into account China's natural conditions and characteristics, rather than wholesale copying of foreign patterns and practices.

Based on this rationale, at the 3rd Session of the 18th CPC Central Committee it was announced that China "will be committed to pursuing new-type urbanization with Chinese characteristics, putting the people and their interests first, promoting coordinated development cities of different sizes and integrated development of industries and cities, and the establishment of new rural areas." This was the first time that the CPC Central Committee mentioned "the path of a new type of urbanization with Chinese characteristics" in an official document. Such a path is indeed an approach to urbanization that is based upon conditions in China, that meets the criteria in the Scientific Outlook on Development, and that is diversified, progressive, effective, harmonious and sustainable. It not only reflects Chinese characteristics, but is also an approach that meets the characteristics of new urbanization, including a people-centric focus, high efficiency and intelligence, green development and low carbon emissions, integrated urban and rural development, and coordination of industrialization, urbanization, IT application and agricultural modernization. These features are an important part of socialism with Chinese characteristics (Wei 2014).

At the Central Work Conference on Urbanization in December 2013, it was emphasized that at the core of the path to new urbanization with Chinese characteristics and scientifically-sound development is the principle of putting the people and their interests first. Moreover, the key to a people-oriented approach was to improve the quality of urbanization and coordination with industrialization, IT application and agricultural modernization. This idea was further clarified in the *National New Urbanization Plan* (2014–2020) released on March 16, 2014. This plan noted, "Urbanization in China is carried out within the context of China's primary characteristics of a huge population, a relative shortage of resources, a vulnerable environment and unbalanced urban and rural development. Urbanization in China must be based upon the primary fact that China is still in the early stage of socialism, it must follow the universal laws of urbanization, and it must focus on being people-first, proper coordination of urbanization, industrialization, IT application and agricultural modernization, optimal deployment, eco-friendliness and the perpetuation of cultural heritage." This is the official definition of the path to a new type of urbanization with Chinese characteristics.

Generally, Chinese scholars' understanding of urbanization with Chinese characteristics and a new urbanization has been evolving from an approach that separates the two concepts completely, toward viewing them as an organic whole. This evolution suggests that Chinese scholars are deepening their studies and reaching consensus about the path to a new type of urbanization with Chinese characteristics. The government's understanding of urbanization has also been evolving along these lines. China is undoubtedly in a critical period of economic transformation and continuing urbanization. If China continues on the traditional, ineffective pattern of urbanization, it will face challenges such as slow industrial structure improvement, resource shortages and environmental degradation, and increasing social conflicts. Chinese scholars and the government are in consensus that China should adhere to a path of new urbanization with Chinese characteristics that focuses on the quality of urbanization.

2.2 The Scientific Import of New Urbanization with Chinese Characteristics

The concept of the path to new urbanization with Chinese characteristics is an organic combination of the concepts of urbanization with Chinese characteristics and new urbanization. As mentioned above, this new approach is one that meets China's natural conditions and adheres to the criteria laid out in the Scientific Outlook on Development, with emphasis on a people-centric focus, high efficiency and intelligence, green development and low carbon emissions, integrated urban and rural development, coordination among industrialization, urbanization, IT application and agricultural modernization, and cultural inheritance. These features are an important part of socialism with Chinese characteristics. The people-centric focus is at the core of, and the essential requirement for, new urbanization with Chinese characteristics. In order to carry out new urbanization with Chinese characteristics, people must be placed at the core of the urbanization process, with the aim of increasing the well-being of the people and prioritizing public needs and interests. High efficiency and intelligence, along with green development and low carbon emissions, are also among the major characteristics of new urbanization with Chinese characteristics. New patterns and methods must be adopted to transform urbanization in China so that they have these characteristics. Lastly, integrated urban and rural development, and coordination among industrialization, urbanization, IT application, agricultural modernization and cultural inheritance constitute another major foundation of new urbanization with Chinese characteristics. Urbanization is not just about building towns and cities, but also involves urban and rural development and prosperity built upon the above factors.

2.2.1 Putting the People and Their Interests First

New urbanization with Chinese characteristics must put the people and their interests first, with the aim of improving the well-being of urban and rural residents. China should pay more attention to people's livelihood and social issues and accelerate the hukou urbanization of rural migrants so as to facilitate equal access to opportunities and benefits for urban and rural residents alike. This will put China on the path to harmonious urbanization characterized by equality, inclusiveness and safety. In order to accelerate the hukou urbanization of rural migrants, China should first work to eliminate the discrimination against people with rural *hukou*, and assist rural migrants in accessing basic urban public services such as childhood education, employment, basic elderly care, basic medical care and affordable housing. Second, rural migrants into cities should be granted the same rights and benefits available to urban residents, including more rights to property. Migrating into a city should not be at the expense of losing one's rights to property in his/her rural hometown. Such property includes farmhouses, curtilage, and the right to contract land and shares in collectively-owned assets. As a growing number of people move into cities from the countryside, a decrease in the number of farmers and improved property rights transactions will form an integral component of a new, people-centric pattern of urbanization. These changes will increase farmers' incomes, fuel the growth of large-scale agricultural production and stimulate rural economic growth, thereby addressing the Three Rural Issues—agriculture, rural areas and farmers. Lastly, China should increase efforts to improve rural migrants' scientific and cultural literacy so as to help them gain a better understanding of urban civilization, lifestyles and values, thereby facilitating healthy urbanization.

2.2.2 *Efficient and Smart*

New urbanization with Chinese characteristics must be based upon China's natural conditions and guided by the Scientific Outlook on Development. To achieve high-quality, health urbanization, measures should be taken to transform the focus of urbanization in China from quantity toward quality, from low efficiency toward high efficiency and from unsustainability toward sustainability. This can be accomplished through adherence to the basic state policy of resource conservation, the wide use of energy-saving technologies and materials to reduce water and land consumption in cities, the construction of energy- and land-efficient buildings, and the development of resource-efficient patterns of production, living and consumption. There are several distinct steps that can be taken to achieve highly effective urbanization. The first is building long-term mechanisms for utilizing resources in an effective and efficient manner. Adopting a path to urbanization characterized by efficient and effective use of resources, including land, will reduce resource consumption in the process of urbanization while increasing the allocative efficiency of urban resources. The second is determining the right building density in towns and cities. It is necessary to study effectiveness metrics and building standards and emphasize a pattern of urban development characterized by compactness, centralization and high efficiency. Measures should be taken to unleash the potential of urban land through effective and efficient use. Local governments should be urged to prioritize effectiveness in their development plans, with a focus on efficient land use and the development of compact towns, cities, communities, industrial parks and villages so as to prevent urban sprawl. The third step is to emphasize intelligence in urbanization by combining urban development with the use of intelligent systems. China should make full use of modern science and technology to combine urbanization with IT application and to accelerate the development of intelligent cities, communities and industrial parks. China should work to improve intelligent industrial and transport systems, enhance intelligent city management and rely on intelligent technologies and management to solve "urban diseases." In brief, China should promote intelligent urbanization.

2.2.3 *Green Development and Low Carbon Emissions*

New urbanization with Chinese characteristics must fully embody the concept of eco-friendliness in the process of urbanization. China should facilitate the formation of green, low-carbon patterns of production, living and urban development. First, it is important to obey the laws of nature and always prioritize environmental protection. It is advisable to build open urban ecological corridors and networks that make full use of China's natural topography of mountains, rivers, lakes, forests, farmland, and so forth. Second, China should determine appropriate development intensity and define ecological red lines. It is necessary to reasonably balance production, living and ecological spaces to build sustainable, livable, beautiful towns and cities. The report of the Central Urbanization Work Conference emphasized that China should integrate cities with nature so that residents have views of mountains and rivers/lakes. Third, China should emphasize green development characterized by a circular economy and low carbon emissions. A component of green development is the efficient and effective use of resources such as land, water and energy and the enhancement of environmental protection and ecological restoration so as to reduce disturbance and damage to nature. Fourth, China should broaden the use of energy-efficient, green, low-carbon technologies and accelerate the building of green production and consumption systems. Cities should be encouraged to develop so that they become eco-friendly, garden-like and/or forest-rich. Measures should be taken to establish a pattern of urbanization compatible with the region's resource carrying capacity and the larger environment. Urban development should be organically combined with environmental protection to achieve sustainable urbanization characterized by green development, low carbon emissions, environmental friendliness and livability.

2.2.4 Integrated Urban and Rural Development

New urbanization with Chinese characteristics must eliminate the urban-rural divide and facilitate interaction and integration between urban and rural areas. It is necessary to actively explore a new pattern of urbanization that enables integrated urban and rural development rather than just focus on urban development. There are several steps needed to accomplish this. First, China must make a uniform plan for urban and rural development. This requires thinking beyond the boundaries of urban and rural areas and embracing the idea of developing them under a single, overall plan. To this end, it is necessary to view cities and the countryside as an organic whole, establish a comprehensive plan, and promote the integration of urban and rural planning and management. Second, China must promote the integration of urban and rural infrastructures. A single, overall plan is needed to enhance the construction of rural infrastructure and facilitate the extension of urban infrastructure into rural areas to build an integrated infrastructure network in both rural and urban areas. Third, China must promote the integration of industrial development in urban and rural areas. This requires the elimination of institutions and systems that have caused the urban-rural divide and a holistic view of industries in urban and rural areas for overall planning. It is necessary to integrate urban and rural resources, create industrial ties between urban and rural areas, facilitate the flow of the factors of production from cities into the countryside and guide the expansion of urban industries and enterprises into the countryside. This will facilitate the symbiosis of urban and rural industries characterized by mutual support and integration. Fourth, China should promote the integration of urban and rural public services. This requires accelerating the construction of the rural public service system and facilitating the extension of urban public services into the countryside, thereby achieving equal access to basic public services and narrowing the urban and rural gap in public services. Fifth, China should promote the integration of urban and rural environmental protection. This requires a shift in thinking in which rural environmental protection is viewed as equally important as urban environmental protection, and in which the environmental quality of urban and rural areas is planned, implemented and managed in a uniform manner. Sixth, China must promote the integration of urban and rural social governance. This requires a single, overall plan for urban and rural social governance, improvements to the means of social governance and social governance service systems, acceleration of the integration of urban and rural societies, and improved means to address social conflicts and threats to social harmony in a timely manner. In brief, China should work to establish a new pattern of integrated social governance in urban and rural areas.

2.2.5 Coordinating Industrialization, Urbanization, IT Application and Agricultural Modernization

New urbanization with Chinese characteristics must facilitate the integration of IT application and industrialization, enhance interaction between industrialization and urbanization, and improve coordination between urbanization and agricultural modernization. A simple way of viewing these four concepts is viewing industrialization as a support, urbanization as a carrier, IT application as a means and agricultural modernization as the foundation. Urbanization must be closely combined with the other three concepts to achieve a new pattern of coordinated development. The first aspect is integrating IT application with industrialization. China should aggressively develop electronic, IT and other high technologies and improve IT application in traditional industries so as to promote industrial structure improvement and increase the level of industrialization. The second aspect is achieving positive interaction between industrialization and urbanization. On one side, urbanization plays a significant role in stimulating investment and domestic demand so as to create market demand and room for industrialization. On the other side, the positive effects of industrialization on industrial development and job creation can expand and improve urbanization. Promoting interaction between industrialization and urbanization will create opportunities for industrialization and power urbanization. The third aspect is coordinating urbanization and agricultural modernization. On the one hand, China should promote urbanization in order to create jobs for rural migrants, modernize agricultural equipment and increase agricultural productivity while creating conditions and markets for agricultural modernization. On the other hand, promoting agricultural modernization and strengthening agriculture will help ensure China's food security, thereby supporting urbanization. Urbanization and agricultural modernization are mutually supportive, and coordination between the two is helpful in addressing "urban diseases" and "rural diseases," and ultimately will improve urban and rural prosperity together.

2.2.6 Perpetuation of Cultural Heritage

New urbanization with Chinese characteristics must reflect China's history and culture. Practical, distinctive patterns or urbanization should be cultivated to build beautiful towns and cities that reflect historic character, cultural heritage, local features and ethnic characteristics. Instead of copying foreign architecture, the new China urbanization model should strive to highlight local features and ethnic characteristics and incorporate local culture, Chinese features, and cultural characteristics into town and city development. It is advisable to underscore natural, historical and cultural elements of the region, embody regional differences and advocate diversity. While China should seek to prevent cookie-cutter cities and buildings, it should also seek to temper the current obsession with weird buildings. Measures should be taken to protect and retain local cultural traditions and historical significance of objects, buildings, streets and blocks during the process of urban reconstruction. China should also focus on building distinctive, beautiful villages as part of building a new socialist countryside. This is an important component of new urbanization with Chinese characteristics. With a focus on improving the lives of farmers, construction of new villages should enhance rural infrastructure and service networks and functionally improve natural villages for the convenience of farmers at work and in life. However, it is also important to maintain the original features and traditional cultures of villages, especially ethnic minority villages and buildings with historical, artistic and scientific value.

In sum, new urbanization with Chinese characteristics refers to sustainable urbanization that is guided by the Chinese government's Scientific Outlook on Development, is based upon China's basic characteristics (i.e., large population, a relative shortage of land and other resources and wide urban-rural differences), emphasizes a people-centric approach, high effectiveness and intelligence, prioritizes green development and low carbon emissions, integrates urban and rural development, embraces full coordination of industrialization, urbanization, IT application, agricultural modernization and cultural inheritance, and is characterized by diversity, progressiveness, high effectiveness, harmony and sustainability. These factors will lead to a new landscape of urban development characterized by resources conservation, eco-friendliness, economic efficiency and social harmony.

2.3 Basic Features of New Urbanization with Chinese Characteristics

The basic characteristics of new urbanization with Chinese characteristics are diversity, progressiveness, high effectiveness, harmony and sustainability.

Diversity

China is a diverse country with a plethora of ethnic groups, cultures, socioeconomic conditions and development levels. Future urbanization in China should account for China's diverse conditions and be carried out in diverse, locally specific manners.

First, China's diverse socioeconomic development levels have led to the existence of multiple stages of urbanization. A large country like China has very big interregional differences with regard to socioeconomic development conditions, and as a result, has varying phases and levels of urbanization. In 2013, China had an average urbanization rate of 53.73%, suggesting that urbanization was slowing down. Yet there were extremely big differences in urbanization rate among provinces, autonomous regions and municipalities directly under the Central Government. Specifically, the urbanization rate was over 80% in the cities of Beijing, Tianjin and Shanghai and over 60% in the provinces of Guangdong, Liaoning, Zhejiang, Jiangsu and Fujian. In contrast, it was below 30% in Tibet and below 40% in Guizhou Province (see Fig. 2.1). Beijing, Tianjin and Shanghai, in particular, were already in the late stage of urbanization. Tibet remained in the early stage, and the other provinces were all in the intermediate stage.

Secondly, China has a diversity of city sizes ranging from towns to small and medium-sized cities to large cities. Because of China's vast size, population, and varying regional conditions, it is important to have both large and small cities. On one hand, the building of large, very large and even mega-cities will produce significant ripple effects from the center outward. On the other hand, it is important to have a large number of specialized towns and small and medium-sized cities to serve as bridges given their lower entry barriers and close ties with rural areas. Coordinated development of both large and small cities will lead to a network of towns and cities of varying sizes characterized by reasonable division of labor and access.

Third, China has a diversity of region-specific patterns of urbanization. Chinese regions should adopt different urbanization strategies based on their specific development stages and conditions. Regions like the Pearl River Delta, the Yangtze River Delta, the Beijing-Tianjin-Hebei region and the middle reaches of the Yangtze River should focus on building metropolitan areas and urban agglomerations in order to increase sustainability and promote regional economic integration. Underdeveloped regions such as southwestern and northwestern China should focus on developing central towns and cities. Mountainous regions such as the provinces of Yunnan and Guizhou must explore a mountain-specific pattern of urbanization rather than copy the patterns of other regions such as the Pearl and Yangtze River Deltas, which have different development and regional conditions.

Fourth, there is a diversity of forces driving urbanization. One driver is the government-market relationship. This relationship prioritizes the removal of the

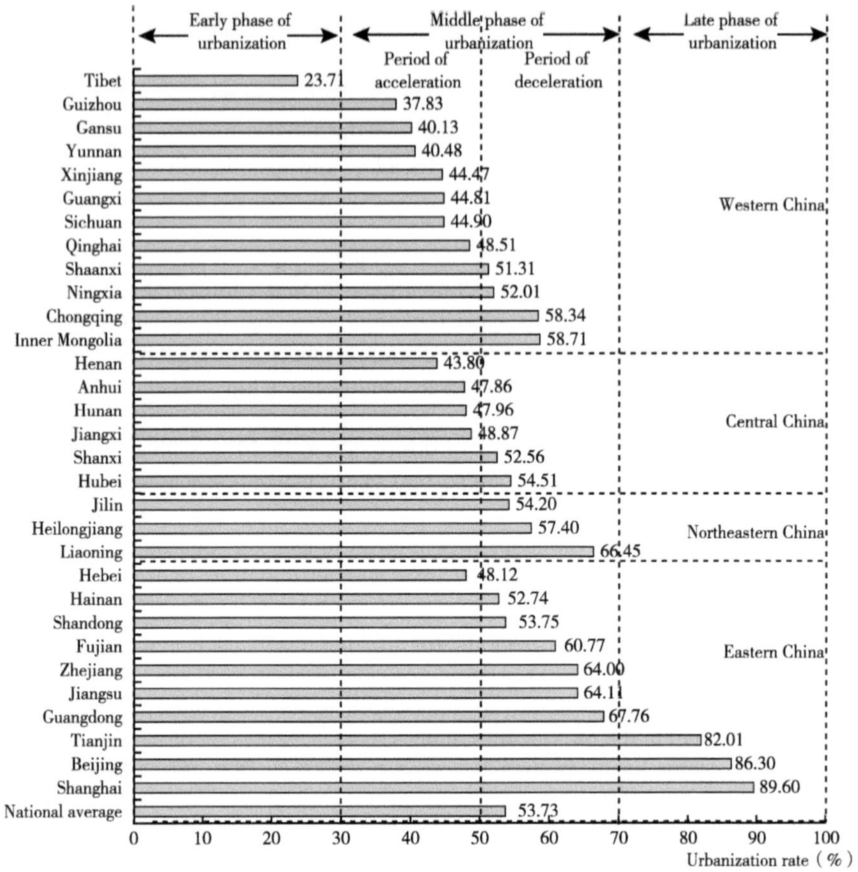

Fig. 2.1 Urbanization rates across China, 2013. *Source* China Statistical Abstract (2014)

urban-rural divide and regional barriers in order to promote the free flow of people and the factors of production to allow the market to play a leading role in resource allocation while the government plays a guiding and planning role. Another driver is the need for industrial support. It is necessary to achieve interaction between industrialization and urbanization and aggressively develop the tertiary sector so as to fuel urbanization through the growth of the secondary and tertiary sectors. Public and non-public economies are also important driving forces for urbanization, as are farmers. City-life is very attractive to farmers, and such attraction is another driving force for urbanization.

Progressiveness

Since urbanization is a long process, China must achieve urbanization in a progressive manner. To this end, regional urbanization must be based upon local conditions and rural migration into towns and cities must be guided at a controlled, reasonable rate.

The rate of urbanization should be commensurate with the level of socioeconomic development in each region and with each town or city's overall carrying capacity and population capacity. It is important to take measures to prevent urbanization from occurring at a rate that exceeds the carrying capacity and the level of socioeconomic development. A big lesson in this regard can be drawn from China's experience with the Great Leap Forward campaign.

First, urbanization must occur at a reasonable rate and degree because of the basic conditions in China, which include a large number of rural residents, scarcity of farmland, heavy pressure to create jobs in urban areas and the relatively insufficient carrying capacity of resources and the environment. Considerations in this regard include the financial strength and the level of economic development in China, the overall carrying capacity, and the capacity of each town or city to accommodate migrants. Both investments into urbanization and the spatial expansion of towns and cities should be kept within a reasonable range. Urbanization should not be carried out at the expense of farmland, food supply and agriculture. Instead, urbanization should be organically combined with agricultural modernization and the building of a new countryside.

Second, the level of urbanization should match that of economic development. In most countries and regions, the level of urbanization is highly correlated with the level of economic development. Urbanization that is significantly behind or ahead of economic development will lead to a series of socioeconomic problems, as witnessed in the early stage of China's reform and opening up when China's urbanization lagged far behind industrialization and economic development. That gap has significantly narrowed due to fast urbanization in recent years. In 2012, for example, the gross national income (GNI) per capita in China equaled only 57.3% of the global average (The World Bank 2013), but the urbanization rate had already reached the global average at 52.6% (UNDP 2013). Despite this catch-up, fast urbanization in China over the past years has created a situation in which rural migrants into cities have not yet become registered city residents. It will take many years for the process of hukou urbanization to finish.

Third, the degree of urbanization should suit the overall carrying capacity, and the capacity of towns and cities to accommodate migrants. The overall carrying capacity of towns and cities includes the carrying capacity of resources, the environment and urban infrastructure. The capacity of towns and cities to accommodate migrants relies on their capacity to offer jobs and industrial support. For the degree of urbanization to be considered reasonable, both the overall carrying capacity and the capacity to accommodate migrants must be balanced. Healthy urbanization is less likely if either capacity is insufficient. Today, China is a world leader in terms of manufacturing; it is the world's No. 1 producer in a lot of segments. However, this status also means that China's industry, especially its manufacturing industry, is less able to offer as many jobs as before. Moving forward, Chinese towns and cities will need to rely more heavily on the tertiary sector to create jobs for migrants. Accordingly, further urbanization must suit the needs for China's economic transformation and industrial structure improvement. It should take care to avoid exceeding what the economy, and industry in particular, is able to support. Employment has been a big challenge in Chinese cities over the past few years, suggesting that urbanization has exceeded what China's industry can support (Lu 2007).

Effectiveness

Given the average amounts of farmland and other resources per capita, China is facing an increasingly tight supply of resources for economic development and urbanization. China should therefore actively explore how to utilize resources efficiently and effectively and conduct urbanization in a way that minimizes consumption of land and other resources, thereby increasing the allocative efficiency of urban resources.

To do this, China must first use land efficiently and effectively. Given China's huge population, land, especially farmland, is extremely precious in China. Although the output per unit of land in Chinese towns and cities has been increasing rapidly in recent years, land per capita in built-up areas has been increasing even faster. Land has been used in an inefficient and ineffective manner, resulting in severe wastage. The governments of towns and cities rely heavily on land-based revenues and are therefore obsessed with expanding town/city sizes. An unreasonably large amount of land has been used to build development areas, for example. Developers usually stock up on land, resulting in land urbanization progressing faster than population urbanization, and leading to a rapid decrease in farmland as a resource. In order to guarantee the 120 million-hectare red-line identified as the baseline for China's food security, the wastage of land during urbanization must be halted through more efficient and effective land use.

Second, China must utilize other resources efficiently and effectively. Today, fast urbanization in China is based upon an inefficient pattern of development reliant on the high consumption of resources—including energy sources—and heavy investment. Insufficient use of high technology has caused inefficient use of resources and severe wastage. China is already extremely lacking in most natural resources; it has no more than half of the global average in terms of farmland, forest, natural grassland and water resources per capita, not to mention the average recoverable reserves of important mineral resources such as oil, natural gas, coal as well as iron, cooper and aluminum ores per capita. Facing such limited resources, China must emphasize the efficient and effective use of resources, accelerate urban development characterized by the conservation of energy, land, water and materials and adopt a resource-saving approach to urbanization.

Third, China must promote compact urbanization. Compact urbanization refers to determining the right building density of various towns and cities, develop effectiveness metrics and building standards, prioritize public transport, and encourage low-carbon mobility. China should advocate the use of a single piece of land for multiple purposes, optimize the structure of urban spaces and increase the accessibility of public infrastructure so as to reduce the time spent on the road. Promoting the efficient use of land for urban development and building compact towns and cities, communities, industrial parks and villages will also help alleviate urban sprawl and preserve societal and cultural diversity.

Fourth, China should promote the co-agglomeration of people and industry, so that cities become the intersection where people, the factors of production and industries meet. During the process of urbanization in China in past years, local governments worked diligently to attract investment and create industry clusters, but were less willing to accept migrants. As an example, some local governments only welcomed land development and other needed resources, and nobody but investors and a limited number of workers were welcomed into the city. This resulted in a severe mismatch between population and industry distributions. To address this issue, local governments must accept the concept of "coordinated development," rely on industry clustering to attract people and promote a co-agglomeration of people and industry with the carrying capacity of local resources and the environment in mind. They should work to ensure a match between population and industry distributions and between the population/economy and the carrying capacity of local resources and the environment, thereby maximizing the co-agglomeration effects of towns and cities.

Harmony

The current status of urbanization in China faces significant problems including the lack of coordination and non-inclusiveness, which threatens to induce social issues and hinder the building of a harmonious society. The process of urbanization involves relationships between urban and rural areas, native residents and new migrants, and the economy and society. The government must effectively handle all such relations, alleviate relevant conflicts and promote equal access to opportunities and sharing of benefits between urban and rural residents. In brief, the government must adopt an approach to harmonious urbanization characterized by equality, inclusiveness and safety.

The first step to achieving more harmonious urbanization is equal access to opportunities. China has long had a *hukou* system that has created an urban-rural divide, characterized by a lack of equal access to development opportunities between urban and rural residents. The restrictive *hukou* system up to today still makes it impossible for rural migrant workers to receive the same treatment as native urban residents in terms of domestic rights, employment, childhood education, social security and car/house purchasing. To provide equal access to development opportunities, therefore, the Central Government must accelerate reform of the *hukou* system, prohibit local governments from issuing any new policies linked with *hukou* types and identify such existing policies. Discriminatory standards based on type of *hukou* must be canceled and existing policies be changed so that they are no longer linked with one's *hukou*. In so doing, citizens' rights and benefits will gradually become independent of the *hukou* system, thus achieving equality in terms of social status and rights (Wei 2013). Accordingly, rural migrants into cities will gain equal access to development opportunities, enabling them to compete with others in the same environment and under the same rules.

The second aspect is benefits sharing. There is now a wide gap between urban and rural residents in China in terms of income and public services. In addition, there is a widening gap in income, social class and living space among urban residents, and rural migrant workers find it difficult to assert their rights and benefits. Such a situation is inconsistent with the essence of socialism. For urbanization benefits to be shared by all urban and rural residents, the Chinese government must accelerate the hukou urbanization of rural migrant workers and cover all permanent residents with basic urban public services. At the same time, it must increase efforts to reconstruct villages in the city, shanty towns and marginalized areas, pay more attention to the urban poor and low-income groups and eliminate the new dual structure in cities. On this basis, the Chinese government should build a uniform urban and rural social security system and a public service system that ensures equal access, covering all permanent residents, both urban and rural, with basic public services. From a long-term perspective, it is necessary to achieve equal access to basic public services throughout China so that all residents—whether urban or rural, in eastern or western regions—will receive consistent, compulsory education, basic medical and health services, social security, safety, and so forth. In addition, the Chinese government should enable all urban and rural residents to enjoy close to the same standard of living and make urbanization benefits available to the general public.

The third aspect to harmonious urbanization is assurance of safety. Assuring safety is an important part of new urbanization given high population densities, the concentration of economic activities and vulnerable ecosystems in cities. First, it is necessary to enhance the capacity of towns and cities in preventing and mitigating disasters. To this end, local governments should accelerate the development of urban disaster reduction and prevention systems, enhance the overall capacity of disaster assessment, make and implement contingency plans, conduct disaster-prevention exercises, build relevant infrastructure, compile stores of relief supplies and disaster kits, and raise urban awareness and response to prevent and mitigate disasters and the loss of life in the event of disasters. Second, it is necessary to ensure residents' safety at work. More attention should be paid to industries where dangers like poisoning are more likely to occur, such as coal mining, chemical manufacturing, smelting and firecracker production. Third, it is necessary to enhance the overall safety of urban areas. Efforts should be made to enhance urban infrastructure, improve legal and regulatory systems, and reduce the occurrence of accidents with respect to food and driving safety resulting from the malfunctioning of urban systems.

Sustainability

Future urbanization in China must not be carried out at the expense of farmland availability, food supply and the environment. Instead, it must focus on quality with greater emphasis on farmland and environmental protection and the building of green production and consumption systems. Measures should be taken to promote the deep integration of urban economic development with environmental protection. This will enable the development of beautiful towns and cities characterized by green development, low carbon emissions, environmental friendliness and livableness.

The first step toward sustainability is coordinating urbanization with agricultural modernization. During fast urbanization, the spatial expansion of towns and cities typically entails a significant loss in farmland. Although the Chinese government has implemented a policy aimed at making up for lost farmland by offering reclaimed land with the same area and quality, the incontestable reality is that the reclaimed land is of lower quality than the lost farmland. In the process of urbanization, therefore, the government must establish and adhere to the most stringent farmland protection and compensation systems to ensure that the used farmland is made up for by reclaimed land with the same area and quality. Mechanisms for protecting basic farmland, in particular, must be created in accordance with relevant objectives in order to fully improve the protection of farmland that ensures food supply in China. In brief, greater effort should be made to ensure China's food security and further coordinate urbanization with farmland protection and agricultural modernization.

The second aspect of sustainability is coordination between urbanization and environmental protection. The fast spatial expansion of urban areas amid urbanization has caused the disappearance of farmland and natural ecosystems like water bodies. It has also led to a rapid decrease in wetland area and biodiversity. Moreover, the sprawl of roads, squares, public facilities and other man-made structures is turning some urbanized areas into concrete jungles. For future urbanization, the government must pay much attention to ecological restoration and environmental protection. Open urban ecological corridors and networks that integrate mountains, rivers, lakes, forests, farmland, and so forth into the urbanization plan should be built. At the same time, the government should focus on developing cleaner production and a circular economy to reduce waste and enhance environmental governance. It is necessary to coordinate urbanization with environmental protection in order to achieve sustainable urbanization.

The third aspect to achieving sustainability is to build sustainable, livable, beautiful towns and cities. Urban areas are not only spaces where economic activities occur, but are also spaces for living and recreation. It is necessary to maintain a reasonable proportion of spaces for production, living and ecological purposes. In the age of the city, urban residents will place more emphasis on quality of life and improving the living environment. This will lead to greater demand for better recreational, public and green spaces in urban areas. Accordingly, in the transition toward green urban development, sustainable and livable towns and cities consisting of eco-friendly buildings, industrial parks and communities, and a good living environment will become an important part of sustainable urbanization.

References

Gao, X. C., Zhou, Y., & Xu, J. (2010). On the historical necessity of adopting an approach to urbanization with Chinese characteristics. *Productivity Research, 1,* 15–16.

Jian, X. H. (2003). On an approach to urbanization with Chinese characteristics—Part 2 of study on an approach to urbanization with Chinese characteristics. *Study and Practice, 11,* 45–51.

Lu, D. D. (2007). Urbanization in China: Process and spatial expansion. *China Urban Economy, 10,* 14–17.

Luo, H. B. (2010, February 20). On the meaning and significance of "a new type of urbanization". *Hunan Daily*.

Ma, K. (2012). Change the pattern of urbanization and improve its quality on the road to China-model urbanization. *Journal of Chinese Academy of Governance, 5,* 4–12.

Ni, P. F. (2013). On a new type of urbanization: the basic pattern of, specific path and promotive measures. *Jianghai Academic Journal, 1,* 87–94.

Niu, F. R. (2010). On the road to urbanization with Chinese characteristics. *Journal of Theoretical Reference, 2,* 6–9.

Qiu, B. X. (2010). Make a scientifically-sound plan and carefully implement the strategy of new-type urbanization. *Planners, 26*(7), 7–12.

Qiu, B. X. (2012). A new type of urbanization: From concept to action. *Administration Reform, 11,* 11–18.

The World Bank. (2013). *World Development Report 2014: Risk and Opportunity—Managing Risk for Development*. Washington DC.

UNDP (2013). *Human Development Reports 2013*. The Rise of the South: Human Progress in a Diverse World. New York.

Wang, Y. S., Ou, J. Z., & Hou, S. (2011, October 14). Report on well-coordinated and scientifically-sound urbanization, industrialization and agricultural modernization led by a new type of urbanization. *Henan Daily*.

Wei, H. K. (2013). Accelerate reforming of the *hukou* system: Ideas and measures. *China Development Observation, 3,* 15–17.

Wei, H. K. (2014). *A new road to urbanization with Chinese characteristics*. Beijing: Social Sciences Academic Press (China).

Wei, H. K. (2010, September 16). On diversification: An approach to China-model urbanization. *Chinese Social Sciences Today*.

Wu, J. (2010). On the pattern of urbanization with Chinese characteristics: Issues and thoughts. *Bulletin of Chinese Academy of Sciences, 25*(3), 258–263.

Xiao, J. C., Shi, Y. L., Shen, B., Yuan, Z., Li, Z., & Ouyang, H. (2008). On an approach to urbanization with Chinese characteristics: meaning and path. *Macroeconomic Management, 11,* 19–22.

Zhang, Z. B. (2013). On a new type of urbanization: strategic significance and reform challenges. *Journal of Chinese Academy of Governance, 1,* 48–54.

Zhou, G. Z. (2009). Exploring a road to urbanization with China characteristics. *Urban Planning International*, (supplement), 6–7.

Chapter 3
Strategic Thinking About Urbanization in China

Urbanization in China is currently undergoing a strategic shift away from acceleration to deceleration and from a focus on quantity to a focus on quality. This shift seeks to remedy the lack of control and urban-rural divide that characterized previous stages of urbanization. The goal of this strategic shift is to establish more effective control and achieve integrated urban and rural development and more equal sharing of benefits. This new stage of urbanization is people-centric, focused on the systematic "hukou urbanization" of rural migrants, the development of urban features and quality improvements, and the reduction of resource consumption and environmental impact. China's efforts in this stage should be geared toward promoting integrated urban and rural development and improving the quality of urbanization. At the same, China should also focus on building an integrated transport network and implementing a network development strategy centered on urban agglomerations and central cities in order to build a system of major economic development belts. These belts, led by the concept of "four longitudinal and four latitudinal belts" will help create a scientifically-sound spatial distribution of towns and cities.

3.1 China's Current Transitional Stage of Urbanization

Will urbanization in China continue to accelerate or start to decelerate during the 12th Five-Year Plan period and beyond? This question is a core strategic issue that deserves close study. The prevailing view is that future urbanization in China will continue to accelerate. It is even believed by some that future urbanization in China will reach a climax. Such a view is based upon the S-shaped curve thesis developed by Northam (1979), which shows that the 30–70% range is the stage of accelerated urbanization (Jian and Huang 2010). We believe that this view is based on a theoretical misunderstanding, and that future urbanization in China will enter a stage of deceleration.

H. Wei, *Urbanization in China*, Research Series on the Chinese Dream and China's Development Path, https://doi.org/10.1007/978-981-13-1408-7_3

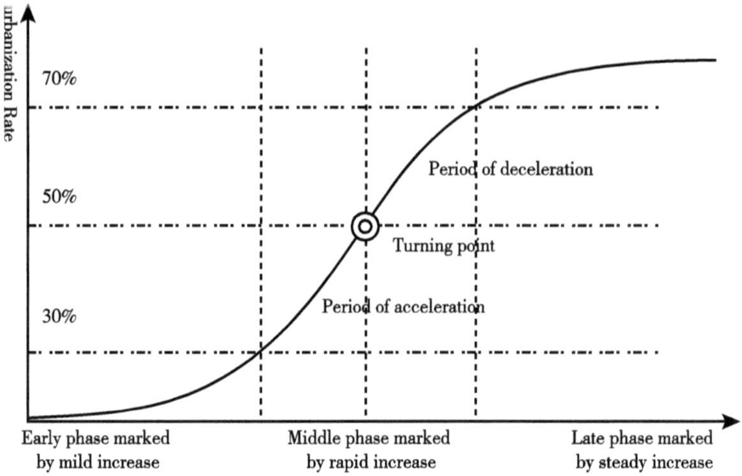

Fig. 3.1 S-shaped curve tracing the trajectory of urbanization process. *Source* The author

First, an urbanization rate range of 30–70% reflects a stage of fast urbanization, but not necessarily accelerating urbanization (Zhou 2005, 2006). Some scholars have used rigorous mathematical models to verify that the maximum acceleration of the urbanization rate at the inflection point of the S-shaped curve is zero, with the rate increasing before this point and decreasing after it (Chen and Zhou 2007; Wang and Wu 2009). Although urbanization occurs at a fast pace in the 30–70% range, the urbanization rate of 50% is an important inflection point. The stage of fast urbanization can be divided at this point into the sub-stages of acceleration and deceleration (see Fig. 3.1). Specifically, the 30–50% range is the sub-stage of acceleration and the 50–70% range is the sub-stage of deceleration. If urbanization continues to accelerate in the 50–70% range, then it compromises the ability of urbanization to move smoothly into the stage of steady development after the urbanization rate is higher than 70%.

Second, according to the history of developed countries, urbanization enters a sub-stage of deceleration after the urbanization rate is higher than 50%. In the United States, for example, the urbanization rate was 28.2% in 1880, 51.2% in 1920 and 69.9% in 1960. The urbanization rate increased by 0.58 percentage points per year on average in the acceleration sub-stage (1880–1920) and 0.47 percentage points per year on average in the deceleration sub-stage (1920–1960). This is a difference of 0.11 percentage points between the two sub-stages (see Fig. 3.2).

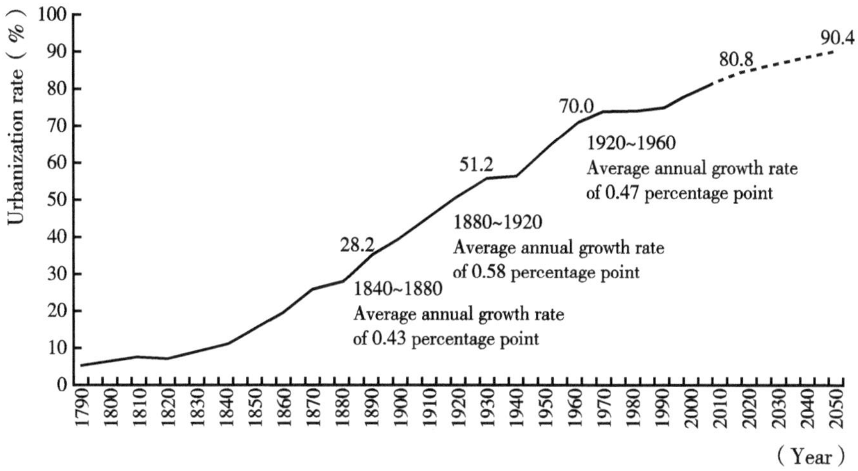

Fig. 3.2 Changes in urbanization rates in the U.S., 1790–2050. *Note* The dotted line is estimates. *Source* Data from the United Nations (2012)

Third, urbanization in China has been showing signs of deceleration in more recent years. The urbanization rate in China increased by 1.44 percentage points per year on average in the 9th Five-Year Plan period, by 1.35 percentage points in the 10th Five-Year Plan period and by 1.39 percentage points in the 11th Five-Year Plan period. Compared to these rates, the growth rate in China has begun to slow down in the 12th Five-Year Plan period, with the urbanization rate increasing by only 1.26% points per year on average in the first three years (see Tables 3.1 and 3.2). The deceleration of urbanization is very obvious in developed eastern regions, in particular. The urbanization rate in Zhejiang Province, for example, increased by 1.37 percentage points per year on average in the 30–50% range (1984–2001), but dropped down to 1.09 percentage points in the period between 2001 and 2013, showing a visible downward trend. Unlike Zhejiang Province, the urbanization rate in Jiangsu Province continued to accelerate after passing the 50% inflection point and did not decelerate until recently. Specifically, the urbanization rate in Jiangsu Province increased by 1.75 percentage points per year on average in the 30–50% range (1993–2005) and by 1.84 percentage points in the period between 2005 and 2012 (see Fig. 3.3). The reason for this is partly because urbanization lagged far behind industrialization in Jiangsu Province due to the reliance on Town and Village Enterprises (TVEs) in the early stages of China's reform and opening up. The adoption of this practice led to industry decentralization with non-urbanization—farmers moved into non-agricultural sectors and worked in factories without leaving their hometowns. As a result, urbanization lagged far behind industrialization in Jiangsu Province after China began reform and opening up. The continued acceleration of urbanization in Jiangsu Province after the 50% inflection point was actually a case of catching up. In 2013, the urbanization rate in Jiangsu Province was 64.11%, with a growth rate already dropping to 1.11 percentage points.

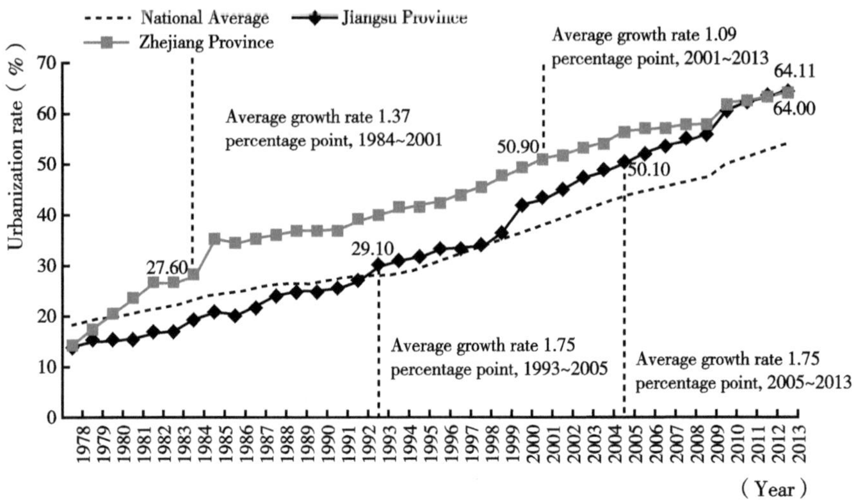

Fig. 3.3 Changes in Urbanization Rates in Zhejiang and Jiangsu, 1978–2013. *Source* China Statistics Yearbook (in previous years), China Statistical Abstract (2014) and statistical yearbooks published by the provinces of Zhejiang and Jiangsu

Statistics show that the overall urbanization rate in China topped 50% for the first time in 2011, reaching 51.27% that year. Based on the points raised above, urbanization in China is now already in the deceleration sub-stage. What this means is that urbanization in China will still proceed at a fast pace in the foreseeable future, but the urbanization rate will increase more slowly. In other words, the rate of growth will decelerate, not accelerate. It is expected that the urbanization rate in China will increase by 0.8–1.0 percentage points per year on average in the foreseeable future, a significantly lower growth rate than the 1.35–1.45 percentage points of the 9th, 10th and 11th Five-Year Plan periods. The high growth rates of these former periods are unlikely to reappear in China, as it would be very difficult to maintain such fast urban expansion.

3.2 Urbanization in China: Trends and Targets

Scholars hold differing opinions on whether the current urbanization rate in China is moderate or not. There are two main viewpoints: urbanization has been going too fast and should be controlled; and urbanization is slow and should become faster. Forecasts for future urbanization trends in China roughly fall under three scenarios (Jian and Huang 2010): urbanization will slow to a growth rate of less than one percentage points per year on average; urbanization will grow at a moderate pace of 1–1.5 percentage points per year on average; and urbanization will continue to grow at a fast rate of over 1.5 percentage points per year on average. Most scholars base their prediction of the future urbanization rate in China upon their judgments of the current urbanization rate. Such forecasts vary widely and even oppose each other. For example, Chen (2000) believes that, given different national conditions, China should not set high urbanization targets like that seen in developed countries but should target an urbanization rate of about 60% by the end of 2050. Wang (2005) believes that, since China's urbanization achievements occurred even with the *hukou* system as a big obstacle, the removal of institutional limitations will accelerate urbanization even further, at a predicted growth rate of 1.44 or higher percentage points per year on average. Other scholars believe that urbanization in China should focus on quality, not merely on rate (Zhou 2005). The rest of this chapter focuses on a discussion of the trends of future urbanization in China overall, and of the four main regions of East, Northeast, Central and West China, in particular, with analysis of relevant targets.

3.2.1 Forecasts for Urbanization Rates in China by the End of 2050

As of 2013, the urbanization rate in China had increased by 35.81 percentage points since China began reform and opening up, or 1.02 percentage points per year on average. In particular, urbanization has occurred at a faster rate since the 9th Five-Year-Plan period, with an annual average urbanization rate increase of 1.37 percentage points. With respect to this level of fast urbanization in recent years, some scholars believe that urbanization in China has departed from the principle of progressiveness and is actually too fast (Lu and Yao 2007). Some believe that the urbanization rate should be controlled and an annual average Urbanization rate increase of 0.6–0.8 percentage points is reasonable (Zhou 2005). This study believes that China will remain in a stage of fast urbanization in the foreseeable future and that a sharp decrease in the urbanization rate from the 1.39 percentage points per year on average in the 11th Five-Year Plan period to 0.6–0.8 percentage points would be unrealistic and unfavorable to China's goal of sustained, steady economic growth. Since the Chinese government views implementation of the urbanization strategy as an important task, urbanization in China will likely continue at a steady, fast rate in the foreseeable future, with an urbanization rate of 1.0 or higher percentage points in particular years.

According to the 2011 Revision of the World Urbanization Prospects, produced by the UN Population Division of the Department of Economic and Social Affairs (DESA), the global rate of urbanization is predicted to grow by 0.39 percentage points per year on average between 2010 and 2050. The urbanization rate is estimated to grow 0.21 percentage points in developed countries, 0.45 percentage points in less developed countries, 0.50 percentage points in Asian countries, 0.46 percentage points in African countries and 0.70 percentage points in China. In China, in particular, the annual average growth of the urbanization rate is predicted to be

Table 3.1 Forecasts for the urbanization rate in China, 2020–2050

Method	2011	2020	2030	2040	2050
UN prediction	49.2[a]	61.0	68.7	73.4	77.3
Experience curve	51.3	59.1	69.5	78.1	85.0
Economic modeling	51.3	61.1	69.6	77.2	84.8
Urban versus rural population growth	51.3	60.8	66.0	70.8	75.1
Comprehensive prediction	**51.3**	**60.3**	**68.4**	**75.4**	**81.6**

Note "a" denotes 2010 data
Source UN forecasts from United Nations (2012) and the others from Wei (2014)

1.18, 0.77, 0.47 and 0.39 percentage points in the periods 2010–2020, 2020–2030, 2030–2040 and 2040–2050 respectively (United Nations 2012).

This study predicts the trends of future urbanization in China using the experience curve, economic modeling, the UN predictions and urban versus rural population growth methods (see Table 3.1). By using a combination of these three methods, this study predicts that the urbanization rate in China will grow by about 1.0 percentage points per year on average in the 12th and 13th Five-Year Plan periods assuming relatively stable national and international environments. A sharp decrease in the urbanization rate will be unlikely, as will the reappearance of the amazingly high growth rate of 1.39 percentage points that occurred since the 9th Five-Year Plan period. This study predicts that urbanization in China will progress at a moderate rate of about 0.8 percentage points per year on average between 2020 and 2030, and will likely slow to 0.6–0.8 percentage points per year on average, between 2030 and 2050. Given these estimates, this study predicts that the urbanization rate in China will reach about 60, 68, 75 and 82% in 2020, 2030, 2040 and 2050 respectively. If the urban versus rural population growth method is used for prediction, then these estimates shift to 61, 66, 71 and 75% respectively.

3.2.2 *Analysis of and Forecasts for Urbanization Rates by Region*

There are obvious differences in the rate of urbanization among Chinese regions today. In 1978, Northeast China had the highest urbanization rate at 37.0%, whereas East, Central and West China had similar urbanization rates of 15.7, 14.1 and 13.8% respectively. However, after 32 years of growth since China's reform and opening, great changes have occurred in these regions, leading to a widening gap in urbanization rate. The eastern regions were the first to benefit from China's reform and opening up. Urbanization in the Pearl and Yangtze River Deltas and the Beijing-Tianjin-Hebei region was put on a fast track thanks to continuous investment by the private sector and foreign companies. The urbanization rate in these regions grew by 2.02 percentage points per year on average in the 6th Five-Year Plan period, 0.89 percentage points in the 7th Five-Year Plan period and over 1.0 percentage point in the 8th, 9th, 10th and 11th Five-Year Plan periods. In 2010, East China surpassed the urbanization rate of Northeast China for the first time in history. With an urbanization rate of 59.7%, East China is China's fastest-urbanizing region.

Northeast China started early in terms of industrialization, and had an urbanization rate of over 37% back in 1978. Afterward, however, urbanization in Northeast China has been progressing slowly with an annual average growth rate of below 1.0 percentage points, making it the slowest-urbanizing region in China. Central and West China had urbanization rates very similar to East China in the early stage of China's reform and opening up. However, both regions experienced a widening rate of urbanization gap with East China, as the latter was the first to benefit from China's reform and opening up. The later implementation of the "Western Development" and "Rise of Central Region" strategies saw a growth in the urbanization rates of Central and West China by over 1.0 percentage points per year on average in the 10th and 11th Five-Year Plan periods. In 2010, the urbanization rate reached 43.6% in Central China and 41.4% in West China. However, the rate of urbanization rate gaps between the Central and Western regions and East China widened to 16.1 and 18.3 percentage points respectively.

Since Chinese regions are in different stages of urbanization, it is clearly impractical to expect all regions to urbanize at the same rate (Zhou 2006). The aforementioned four main regions will differ from one another in terms of urbanization rate and quality due to differences in natural conditions and stages of development. Despite this, this study expects the Urbanization rate gaps among the four main regions will generally narrow. East China will experience a decrease in its urbanization rate within the next few years, and Northeast China will see continued, slow urbanization. Both Central and West China will see higher urbanization rates and thus take the spotlight for urbanization in China. Using the urban versus rural population growth method based on urbanization rate data from 1978–2010, this study predicts the trends of urbanization in the four main regions in 2015, 2020, 2030, 2040 and 2050.

First, the urbanization experience of South Korea and Japan has recurred in China's eastern regions, led by the Pearl and Yangtze River Deltas. The urbanization rate in Guangdong Province grew by over 2.0 percentage points per year on average between 1982 and 2000, while the urbanization rates of Zhejiang and Jiangsu Provinces and Shanghai City grew by over 1.3, 1.4 and 1.5 percentage points respectively (Zhou 2006). That said, the urbanization rate in China's Eastern regions will likely decrease within the next few years. In reality, urbanization in developed eastern regions such as the Yangtze and Pearl River Deltas has already been slowing down. These regions will be the first in China to finish urbanization, with the overall urbanization rate reaching 66.7, 73.0, 77.6 and 81.7% in 2020, 2030, 2040 and 2050 respectively, according to some forecasts (see Fig. 3.4).

Actual Value Estimates

Second, Northeast China will experience continued, slow urbanization. The local Urbanization rate will grow by only 0.6 percentage points per year on average in the

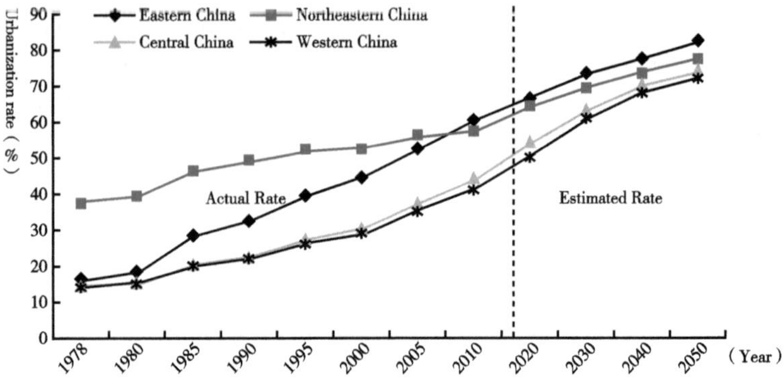

Fig. 3.4 Projected trends in urbanization rate in four major regions in China. *Source* Wei (2014)

12th and 13th Five-Year Plan periods. It will then grow more slowly, with an annual average rate of only about 0.39 percentage points beyond 2030. The urbanization rate in Northeast China will reach 63.6, 69.2, 73.3 and 77.0% in 2020, 2030, 2040 and 2050 respectively.

Third, urbanization in Central and West China will remain in the acceleration sub-stage. The local urbanization rates will still grow at a high annual average rate of about 1.0 percentage points in the 12th and 13th Five-Year Plan periods. The urbanization rates in Central and West China will be higher than those of East and Northeast China beyond 2020. The urbanization rate in Central China will reach 53.5, 63.2, 69.9 and 73.9% in 2020, 2030, 2040 and 2050 respectively, while in West China it will reach 51.4, 61.2, 68.0 and 72.2% respectively.

3.2.3 *Targets for Urbanization in China*

In 2013, the urbanization rate in China already exceeded the global average of 53.0% (UNDP 2014) and, of course, passed the inflection point of 50%. This suggests that urbanization in China is now already in a strategic transition—future urbanization will slow down and enter the deceleration sub-stage. For future urbanization in China, the Chinese government must prioritize quality over rate in order to accomplish high-quality, healthy urbanization.

First, the upper limit of 85% urbanization rate in China is expected to be reached around 2050. The global history of urbanization tells us that the urbanization rate will not grow indefinitely and will ultimately hover somewhere between 80 and 90% in most countries and regions, with the special exceptions of Singapore and Monaco, where the urbanization rate can reach 100%. Since China is the world's most populous country, it would simply be unrealistic for its urbanization rate to reach 100%. First, a populous China must have a certain number of people working in agriculture to ensure food security. This is a necessary requirement for national security. Second, as the standard of living increases, many people will still want to live in the countryside in order to enjoy the rustic landscape consisting of mountains, rivers, lakes, forests, farmland, and so forth. Third, no large and populous country has ever become fully urbanized, constrained by the universal laws of urbanization. This study believes that the upper limit of China's urbanization rate is about 85%.[1] According to our prediction, this limit will be reached around 2050, when the structures of urban and rural populations, land and space in China will stabilize.

Second, China must set targets for high-quality, healthy urbanization. The first target should be a moderate urbanization rate of 0.8–1.0 percentage points in between 2013 and 2030, and a slightly lower rate of 0.6–0.8 percentage points in between 2031 and 2050. The second target should be achieving equal hukou urbanization and urbanization rates. China should make great efforts to complete the hukou urbanization of rural migrants by 2025 so as to synchronize hukou urbanization with urbanization. The third target should be an increasingly reasonable pattern of urbanization. China should focus on creating an urban system characterized by clear division of labor, functional and complementary tiers, and a balanced and reasonable spatial structure characterized by high effectiveness and efficiency. This will allow towns and cities to achieve well-coordinated development. The fourth target should be developing more sustainable towns and cities in which the environmental quality of urban areas improves and the resource and environmental costs of urbanization decreases significantly. This will facilitate sustainable urbanization characterized by low consumption, low emissions, high efficiency and affordability. The fifth target should be developing a new urban-rural relationship featuring the free flow of the factors of production, exchange on an equal basis, and balanced allocation of public resources. This will allow urban areas to have significant ripple effects on rural economies, and facilitate the integrated development of rural areas.

Third, China has developed metrics for new urbanization into 2020. These metrics underscore the key characteristics of new urbanization including a people-centric approach, high effectiveness and intelligence, green development and low carbon emissions. Under the *National New Urbanization Plan* (2014–2020), the urbanization rate of China's permanent residents is predicted to reach about 60% by the end of 2020; the urbanization rate of China's permanent residents already registered under the household registration system in their place of permanent residence is estimated to be about 45%. In addition, this Plan outlines the goals of public transport accounting for 60% of road and rail transport in cities with a population of over one million people, renewable energy accounting for 13% of energy consumption in urban areas, green buildings accounting for 50% of new buildings in urban areas, and the average area of land used for urban development per capita controlled within

[1]A higher urbanization rate is not necessarily better. There is a peak value, or upper limit, for a country's urbanization rate depending on food security, reasonable urban-rural population and space ratios. This peak value for China may be about 85% (Wei 2013). But this peak value may also be about 70%, according to another study [the Department of Services and Management for Migrant Population, the National Health and Family Planning Commission of the People (NHFPC) 2012].

100 m^2 (see Table 3.2). Additionally, the Chinese government has committed to ensuring that, by the end of 2020, about 100 million rural migrants living as permanent residents in cities will get registered under the *hukou* system in their places of permanent residence. The Chinese government has also committed to rebuilding shanty towns and villages in the city for about 100 million people and urbanizing about 100 million people in Central and West China.

3.3 Main Tasks for New Urbanization

New urbanization with Chinese characteristics requires replacing the previous practice of prioritizing rate over quality with equal attention to both rate and quality. In order to change the focus of urbanization from rate to quality, it is necessary to accelerate the process of complete urbanization, and organically combine fast urbanization with quality improvement (Wei 2011). Specifically, there will be six main tasks for further urbanization in China in the foreseeable future.

3.3.1 Phasing in the Hukou Urbanization of Rural Migrants

Steadily promoting the hukou urbanization of rural migrants to address the issue of incomplete urbanization is critical to fully improving the quality of urbanization. The report to the 18th Party Congress noted that: "We should … conduct registration of rural migrant workers as permanent urban residents in an orderly way, and endeavor to ensure that all permanent urban residents have access to basic urban public services." Promoting the hukou urbanization of rural migrants was identified as the primary task among all six of the main tasks for urbanization identified at the Central Urbanization Work Conference in December 2013. There are three sections in the *National New Urbanization Plan* (2014–2020) that explain targets, tasks and measures for systematically promoting the hukou urbanization of rural migrants. New urbanization with Chinese characteristics requires a people-centric approach that follows the principles of "Respect individual decisions, let one choose for oneself, adapt to local conditions, implement hukou urbanization step-by-step, prioritize the hukou urbanization of existing migrants, and prepare to accept new migrants" to steadily phase in the hukou urbanization of rural migrants. It is important to gradually provide rural migrants with the same benefits and services available to native urban residents, including social security, employment and new-job training, public services, affordable housing and childhood education. The government should work to eliminate barriers between urban and rural areas and discrimination against rural migrant workers, implement an employment system to which urban and rural residents have equal access, create a public service system that equally benefits urban and rural areas, and promote the building of uniform social security, welfare and *hukou* management systems for urban and rural areas. In addition, it is necessary

Table 3.2 Main targets for new urbanization in China

Indicator		2012	2020
Urbanization rate	Permanent residents (%)	52.6	About 60
	Permanent residents registered on the *hukou* system in their places of permanent residence (%)	35.3	About 45
Basic public services	Children who stay with rural migrant workers and who receive compulsory education in the host city (%)		≥99
	Unemployed native urban residents, rural migrant workers and people newly available for work who receive free, basic occupational skills training (%)		≥95
	Permanent urban residents covered by basic pension insurance (%)	66.9	≥90
	Permanent urban residents who have basic medical insurance (%)	95	98
	Permanent urban residents who live in affordable housing (%)	12.5	≥23
Infrastructure	Public transport's share in road and rail transport in cities with a population of over one million people (%)	45*	60
	Tap water (%)	81.7	90
	Sewage treatment (%)	87.3	95
	MSW treatment into harmless matter (%)	84.8	95
	Broadband network access from home (Mbps)	4	≥50
	Coverage by integrated service facilities in communities (%)	72.5	100
Resources and the environment	Average land area used for urban development per capita (m^2)		≤100
	Renewable energy's share in energy consumption (%)	8.7	13
	Green buildings' share in new buildings (%)	2	50
	Green space percentage of built-up areas (%)	35.7	38.9
	Cities at the prefecture and higher levels that meet the national standard for air quality (%)	40.9	60

Note "*" denotes 2011 data
Source The National New Urbanization Plan (2014–2020)

to include workplace injury insurance, medical insurance and pension insurance for rural migrant workers into compulsory insurance coverage and increase support for rural migrants with respect to housing, childhood education, health, employment and business startup grants, thereby enabling such migrants to harmoniously integrate into cities and share the benefits of urbanization. At the same time, given the high costs of hukou urbanization, reform of the rural property rights system should be organically combined with the hukou urbanization of rural migrants to encourage farmers to move into cities with their assets and build cost sharing mechanisms between governments, businesses and individuals.

3.3.2 Strengthening the Ability of Towns and Cities to Support Industrial Growth and Overall Carrying Capacity

A necessary driver of economic development is equal emphasis on the secondary and tertiary sectors. This requires accelerating the development of high-end manufacturing and eco-efficient urban industries, vigorously developing modern services and labor-intensive industries, and promoting the deep integration of the manufacturing and services industries. In addition, it is necessary to strengthen industrial support for urbanization and provide sufficient employment opportunities for urban residents to achieve the objectives of full employment and quality of life. The government should take measures to promote both industrial development and human agglomeration. Driving agglomeration of people through industrial agglomeration, and determining reasonable sizes of towns and cities on the basis of human agglomeration, will allow a new development pattern to emerge that integrates people and industries into towns and cities.

Effort should be taken to prevent the rise of cities that lack desirable functions and industrial support, or that have too few residents. It is also necessary to speed up infrastructure construction in such areas as urban transport, utilities, communications, housing, education, science and technology, culture, health, sports and elderly care. Improving public services in urban areas, especially towns and small and medium-sized cities, will enhance the population carrying capacity of urban infrastructure. The governments should guide foreign and private investment into urban infrastructure and public services by establishing a diversity of investment and financing platforms and mechanisms. Greater efforts should be made to address "urban diseases" in mega-cities such as Beijing and Shanghai. Efforts should include establishing and improving a joint air pollution prevention and control mechanism, building an integrated high-speed transport and safety monitoring system, and promoting the movement of people, industries and functions from central areas into outer areas, thereby continuously optimizing the spatial structure of metropolitan areas and strengthening the carrying capacity of their resources and environment.

3.3.3 Reducing Resource Consumption as Part of the Cost of Urbanization

The current urbanization pattern that prioritizes rate over effectiveness, quantity over quality and spatial expansion over internal improvement must be changed. The government should undertake efforts to make urbanization more efficient by promoting new urbanization characterized by low energy consumption, low pollution, low emissions and high efficiency and effectiveness. The first step is to respect the carrying capacity of resources and the environment. Urban population size and development intensity should match the overall regional carrying capacity. Regional economic development should be carried out with the goal of preserving the ecological structure and natural environment of the area, as well as maintaining the quality of the human living environment. Measures must be taken to prevent excessive, ineffective or destructive resource exploitation. The second step is to accelerate industrial structure improvements to make more efficient use of the land while giving top priority to human needs. This requires determining reasonable percentages of land used for living, ecological restoration and production, and controlling the price of land used for cities. The government must set upper limits for the percentage of land used for industrial purposes in cities, while gradually increasing the percentages of land used for urban living and ecological restoration. Increasing the efficiency of land use for industrial purposes can be accomplished via strict accordance with the land listing system of inviting bids and holding auctions for such land. The third step is to emphasize quality in the process of urbanization and economic development. This can be accomplished by fueling the growth of an urban circular economy under the principle of the 4Rs (reduce, reuse, re-manufacture and recycle). This requires increased efficiency of resource utilization and tight control over the environmental costs of economic development. At the same time, it is necessary to promote urban industrial structure improvement, optimize the spatial structure of urban areas and increase overall economic efficiency to achieve an organic system of socioeconomic and environmental effectiveness in towns and cities.

3.3.4 Optimizing City Size and Spatial Structures in the Urbanization Process

The current sizes and spatial structures of urban areas in China today are unreasonable and out of sync with the carrying capacity of resources and the environment. This is reflected in the obvious polarization of China's urban development. High population pressures exist in the main urban areas of some mega-cities, while some small and medium size cities suffer insufficient industrial and population bases and there is an excess of small towns with weak service functions. Moreover, the spatial distribution of towns and cities is unbalanced. Urban population density that is either too high or too low is a growing concern, as is the mismatch of industrial and population distributions. Population-dense towns and cities in eastern regions are facing tightening resource and environmental constraints, whereas the urbanization potential of central and western regions with higher resource carrying capacity has yet to be unleashed. New urbanization with Chinese characteristics should seek to address this imbalance and work to optimize city size and spatial distribution. There are two complementary strategies to accomplish this. First, it is necessary to establish workable standards to categorize city sizes, prioritize urban agglomeration as the primary form of urbanization, optimize the development of large cities and mega-cities, and accelerate the development of small and medium-sized cities, with continued support for the development of towns on a selective basis. Urban agglomerations should be the primary model for city size and structure, featuring reasonable division of labor between towns and cities, well-coordinated development and tiers, and effective restraints to prevent polarization in urban growth. The second strategy is to speed up urbanization in central and western regions by improving the industrial systems in those regions, implementing a multi-center network development strategy, and building a system of major economic development belts known as the "four longitudinal and four latitudinal belts." Efforts should be made to change the spatial structure of urbanization from excessive agglomeration toward balanced urbanization to match the carrying capacity of resources and the environment. This study advises establishing a system of urban agglomerations with regional, national and international significance by the end of 2030 and turning them into the primary destination for rural migrants. At the same time, the government should focus on developing an integrated transport network with urban agglomerations as the carriers and major central cities as the nodes in order to build major economic development belts, including the four latitudinal belts along the Yangtze River (along with the Lianyungang-Lanzhou-Urumqi, Shanghai-Kunming and Qingdao-Xining railways), plus four longitudinal belts along China's coastline (along with the Beijing-Guangzhou, Beijing-Shenzhen and Baotou-Nanning railways) to create an effective, efficient and moderately balanced network.

3.3.5 Improving Urban Management Services

The overall quality of urban management services is low in China, as such services typically receive insufficient attention during urban development. Increasingly serious problems include lack of control over the use of urban spaces, excessive population density, traffic congestion, air and water pollution, frequent public safety/security incidents and undersupply of public services. New urbanization with Chinese characteristics requires sufficient attention to improving urban management services. The first step is to replace specialized management with integrated management. Nowadays, the main problems with urban management in China are fragmentation, lack of cooperation, lack of a global view, functional overlap, managerial ineffectiveness and lack of coordination. It is increasingly difficult for the traditional specialized management model to handle the complexities of managing large systems in cities

(Zhou 2002; Zhai et al. 2011; Song and Tang 2007). Going forward, it is necessary to promote integrated urban management and set up a corresponding integrated management department and coordination mechanism in order to ensure that urban management subsystems run smoothly and are consistent with the overall developmental objective of the urban system. The second step is to replace centralized management with participative management. It is necessary to gradually change the government's "do-it-alone" management model to a model of two-way communication between urban management departments and residents. Organizations such as homeowners' committees, property management service providers and volunteer teams should be utilized and guided in participating in community services and social management, thereby improving participative management in a comprehensive manner. The third step is to pay equal attention to daily and emergency management. On the one hand, it is necessary to enhance daily management services closely tied with peoples' everyday lives, such as city appearance, sanitation, landscaping, administrative law enforcement and fee collection, with a focus on addressing major public concerns. On the other hand, it is important to strengthen the urban emergency management system. This requires greater effort in building emergency response infrastructure such as shelters and medical/health care, emergency logistics and transport systems, establishing disaster monitoring and warning systems, and improving contingency plans for sudden public incidents. Enhancing emergency support systems will help ensure that the urban emergency management system is able to warn against, alleviate and even eliminate sudden events such as natural disasters, accidents and public health and safety/security incidents. The fourth step is to roll out grid-based management to make urban management more efficient. Under the grid-based management model, city managers rely on a uniform urban management and digitization platform to divide the managed urban areas into a grid of cells according to a certain standard, implementing a modern urban management method that is active, granular, dynamic, and can act in real-time in a closed loop. Grid-based management is an important instrument for further modernizing urban management as it reduces urban management costs and provides for much more efficient, democratic and better-regulated urban management.

3.3.6 Enhancing Institutional Innovation for Urbanization

Further urbanization in China requires the elimination of the fragmented urban-rural dual structure and accelerated reform of the *hukou*/land/employment management, public service, administration and other systems by establishing and improving a uniform set of such systems for urban and rural residents. These reforms will contribute to equal access to basic public services and the same quality of life between urban and rural residents. It is necessary to create a uniform trading platform and market for land used for urban and rural development, enabling state-owned land and collectively-owned rural land to be equally traded in terms of rights and pricing. Measures should be taken to improve the basic public service supply management system, increase the availability of education and medical care resources and the coverage of social insurance, and further implement the affordable housing policy that benefits various groups. It is necessary to make and release a scientifically-sound standard for setting up city-administered areas as early as possible, explore a China-typical urban system and establish an administrative division system that meets the requirements for social development.

References

Chen, S. R. (2000). On urbanization in China: Status, problems and prospects. *Urban Problems, 1,* 3–6.
Chen, Y. G., & Zhou, Y. X. (2007). On non-linear dynamic models of the urbanization process. *Acta Scientiarum Naturalium Universitatis Pekinensis, 43*(4), 542–548.
Jian, X. H., & Huang, K. (2010). On urbanization in China: Empirical analysis of its level and rate and prediction of its future. *Economic Research Journal, 3,* 28–39.
Lu, D. D., & Yao, S. M. (2007). Urbanization in China from a scientific perspective. *Human Geography, 4,* 1–6.
Northam, R. M. (1979). *Urban geography* (2nd ed., pp. 65–67). New York: Wiley.
Song, G., & Tang, Q. (2007). On modern cities and their management—A type of open, complex, gigantic systems. *Urban Studies, 14*(2), 66–70.
The Department of Services and Management for Migrant Population, the National Health and Family Planning Commission of the People's Republic of China. (2012). *Report on China's Migrant Population Development 2012.* Beijing: China Population Publishing House.
UNDP. (2014). *Human Development Reports 2014.* Sustaining Human Progress: Reducing Vulnerability and Building Resilience. New York.
United Nations. (2012). *World Urbanization Prospects: The 2011 Revision.* New York.
Wang, D. Y. (2005). On urbanization in China and the resulting challenges. *Economic Review, 1,* 4–8.
Wang, J. J., & Wu, Z. Q. (2009). On the stages of urbanization. *Acta Geographica Sinica, 64*(2), 177–188.
Wei, H. K. (2011). On how to adjust China's urbanization strategy. *China Economic & Trade Herald, 7,* 17–18.
Wei, H. K. (2013). Theoretical reflections on city-based society. *Urban Studies, 20*(5), 24–29.
Wei, H. K. (2014). *A new road to urbanization with Chinese characteristics.* Beijing: Social Sciences Academic Press (China).
Zhai, B. H., Li, C., & Yang, F. (2011). Comprehensive modern urban management: A review of its essence, functions and systems. *Shanghai Urban Management, 6,* 17–21.
Zhou, G. Z. (2002). Cities and their areas—A typical open, complex, gigantic system. *Urban Planning, 26*(2), 7–8.
Zhou, Y. X. (2005). On urbanization rate: It's not "the faster, the better". *Scientific Decision Making, 8,* 30–33.
Zhou, Y. X. (2006). On urbanization rate in China. *Urban Planning, 30*(supplement), 32–36.

Chapter 4
Resizing Towns and Cities in China

The size structure of urbanization in China has steadily improved since the country began reform and opening up. Today, China has an urban system consisting of towns and cities of varying sizes. Nonetheless, there is visible polarization in the urbanization process due to a combination of factors that include traditional ideas of development, biased resource allocation, the polarization effect of the market, the desire of farmers to migrate and regulatory ineffectiveness. To address this issue, the Chinese government should establish a scientifically-sound standard for categorizing city sizes, exert differentiated control over the population sizes of mega-cities, cement the position of urban agglomerations as the primary form of urbanization, and increase the overall carrying capacity of small and medium-sized cities. It should take measures to achieve a size structure of towns and cities characterized by reasonable division of labor, well-coordinated development, and a tiered structure.

4.1 Polarization of Urbanization Across China

The Chinese government has implemented a policy to control the size of large cities and to reasonably develop towns and small and medium-sized cities. In the outline of the 10th Five-Year Plan circulated at the 4th Session of the 9th National People's Congress (NPC) in March 2001, it was announced that the Chinese government would take a path to diversified urbanization that suits China's conditions and that enables towns and cities of varying sizes to develop in a well-coordinated manner. The basic guideline of "developing towns and cities of varying sizes in a well-coordinated manner" has since been reflected in various policy documents. From an urban system perspective, the development of towns and cities of varying sizes in a well-coordinated manner is a general law of urbanization around the world and is a fundamental tenet of new urbanization with Chinese characteristics. Yet, the size

H. Wei, *Urbanization in China*, Research Series on the Chinese Dream and China's Development Path, https://doi.org/10.1007/978-981-13-1408-7_4

structure of towns and cities in China is far from balanced, with obvious polarization existing due to a combination of factors (Wei 2014). Such polarization is manifested in three main respects:

4.1.1 Polarization in the Number of Large Versus Small Cities

The first aspect of polarization within China's urbanization is polarization in the number of large cities versus small cities. Large cities are growing in number and percentage, whereas the number of small cities is decreasing. In 1980, the former National Construction Commission revised the *Interim Urban Planning Regulations on Size Limits* to categorize urban population sizes into four levels—over 1,000,000 people (i.e., very large cities), 500,000–1,000,000 people (i.e., large cities), 200,000–500,000 people (i.e., medium-sized cities) and below 200,000 people (i.e., small cities). Implemented in 1990, the *Urban Planning Law of the People's Republic of China* categorized city size into large, medium and small according to the non-agricultural population in districts and suburbs. However, the newer *Urban and Rural Planning Law of the People's Republic of China*, which took effect in 2008, contains no provisions concerning city size categories. Given the rapid increase in China's urban population in recent years, this study has redefined the city size categories to facilitate analysis (see Table 4.1). All Chinese cities are categorized into five levels based on the city's non-agricultural population: over 2,000,000 people (i.e., ultra-large cities), 1,000,000–2,000,000 people (i.e., very large cities), 500,000–1,000,000 people (i.e., large cities), 200,000–500,000 people (i.e., medium-sized cities) and below 200,000 people (i.e., small cities). As can be seen from the table below, there were 461 cities across China in 1990, including nine ultra-large cities, 22 very-large cities, 28 large cities, 119 medium-sized cities and 283 small cities. By 2000, the number of Chinese cities increased to 665, including 13 ultra-large cities, 25 very large cities, 54 large cities, 220 medium-sized cities and 353 small cities. In 2012, there were 655 Chinese cities, including 26 ultra-large cities, 39 very large cities, 95 large cities, 50 medium-sized cities and 245 small cities.

Larger cities with populations over 500,000 people grew in share from 12.79% in 1990 to 13.83% in 2000, before sharply increasing to 24.42% in 2012. In other words, this share increased by 1.04 and 10.59 percentage points in the 1990–2000 period and 2000–2012 period respectively. Medium-sized cities grew in number with their share increasing from 25.81% in 1990 to 33.08% in 2000 to 38.17% in 2012. In other words, their share increased by 7.27 and 5.09 percentage points in these two periods respectively. In comparison, small cities significantly decreased in number, with their share declining from 61.39% in 1990, to 53.08% in 2000, to 37.40% in 2012. In other words, their share decreased by 8.31 and 15.68 percentage points in these two periods respectively. In brief, larger Chinese cities with populations over 500,000 people saw a substantial increase in share, whereas the share of smaller cities decreased significantly. Meanwhile, the number of small and medium-sized cities decreased to from 573 in 2000 to 495 in 2012. This is partly due to the fact

Table 4.1 Variation of percentages of chinese cities in terms of number and population

Urban population (in million people)	1990			2000			2012		
	City count	Percentage by number (%)	Percentage by population (%)	City count	Percentage by number (%)	Percentage by population (%)	City count	Percentage by number (%)	Percentage by population (%)
>2.0	9	1.95	22.92	13	1.95	22.53	26	3.97	33.13
1.0–2.0	22	4.77	18.74	25	3.76	14.55	39	5.95	15.80
0.5–1.0	28	6.07	12.64	54	8.12	15.54	95	14.50	18.76
0.2–0.5	119	25.81	24.64	220	33.08	28.86	250	38.17	22.84
<0.2	283	61.39	21.07	353	53.08	18.52	245	37.40	9.48
Total	461	100	100	665	100	100	655	100	100

Note Urban population sizes are based on non-agricultural populations
Source: *China Population Statistics Yearbook* and *China Population & Employment Statistics Yearbook*

that the Chinese government has suspended the establishment of new cities in recent years and in part due to the fact that some small, medium and large cities have been upgraded to larger city size categories.

To give a clearer picture of the variation of China's city-size structure, we illustrated how city numbers vary according to city sizes, leading to a pyramid of city sizes (see Fig. 4.1). A pyramid-like distribution of city sizes is a general law, namely, larger cities are smaller in number and vice versa (Zhou 2007). China is no exception in this regard. However, observation of the pyramid structure of city sizes shows that while the structure in 1990 and 2000 was stable—the number of cities decreases level by level according to size—the pyramid structure in 2012 is not "firmly grounded." The number of small cities is smaller than that of medium cities, while the number of very large cities is too big and somewhat out of proportion to the large city tier. The pyramid illustration of city sizes intuitively shows a lack of coordination in China's city-size structure in recent years, namely, the number and percentage of small cities is too low as compared to that of larger cities, especially very large cities.

4.1.2 Polarization of Population Percentages by City Size

China's polarized urbanization is also manifested in the polarization of population percentages in cities of different sizes. Specifically, larger cities account for an increasing percentage of China's urban population compared to smaller cities. This is to some extent due to China's long-standing excessively administrative urban system. A multi-tier urban management model based on administrative levels has caused unfair resource allocation. Larger cities, especially China's capital city and other municipalities directly under the control of the Central Government, provincial capitals, and cities specifically designated in the state plan, receive much more resources and growth opportunities than smaller cities. This has resulted in large-scale human and industrial agglomeration in these privileged cities, consequently resulting in rapid increase in size. An empirical study shows that Chinese cities at the prefecture and higher levels generally are close to their maximum efficiency, and

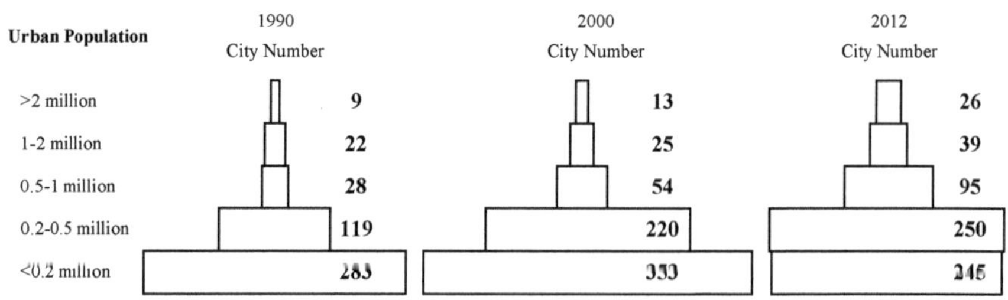

Fig. 4.1 Pyramid illustration of city sizes in China. *Note* Urban population sizes are non-agricultural Source: *China Population Statistics Yearbook* and *China Population & Employment Statistics Yearbook*

many ultra-large cities are already on the decline in this regard (Wang 2010). On the other hand, small and medium-sized cities account for decreasing percentages in total city number and urban population. Some central and western towns and small cities have even shown signs of shrinking due to a lack of strong support from within the urban system.

In this study, we have collected statistics about changes in the total non-agricultural population percentages of cities of different sizes (see Table 4.1). The population percentage of ultra-large large cities increased from 22.92% in 1990 to 33.13% in 2012; that of very large cities increased from 18.74% to 15.80%; that of large cities increased from 12.64% to 18.76%. On the other hand, the population percentages of medium-sized cities decreased from 24.64% to 22.84% and that of small cities decreased from 21.07% to 9.48%, a significant decrease of 11.59 percentage points. In other words, larger cities with populations of over 500,000 people saw an obvious increase in their share of the total Chinese non-agricultural population living in cities, rising a total of 13.39 percentage points in between 1990 and 2012, including a rise of 10.21 percentage points from ultra-large cities. In comparison, small and medium-sized cities with populations of no more than 500,000 people saw a rapid decline in share of population, dropping from 45.71% in 1990 to 32.32% in 2012. This translates to a decline of 13.39 percentage points over 22 years at an annual average reduction of 0.6 percentage points. By the end of 2012, 67.69% of the total non-agricultural urban population lived in cities with populations of over 500,000 people. Of this percentage, cities with populations of over one million people accounted for 48.93% of the population share, and cities with a population of over two million people accounted for 33.13%. These statistics indicate that China's urban population is now highly concentrated in larger cities, resulting in a severely unbalanced population distribution among Chinese cities of different sizes.

In plotting a chart of Chinese cities' population percentages by city size, we see a visual trend that looks very different from the pyramid structure of the distribution of city numbers by size. Rather than accounting for smaller shares of the total number of cities as city size increases (pyramid structure), the population percentages of larger cities in the total population are much higher than their percentages in the total number of cities, which is also a general law worldwide (Zhou 2006). However, Fig. 4.2 shows that the population percentages of Chinese cities of different sizes have been changing from a dumbbell-shaped distribution toward an upside-down pyramid-shaped distribution over the past 22 years. Cities with populations below 200,000 people, in particular, have seen their share drop by 11.6 percentage points. In contrast, cities with populations of over two million people have seen their share rise by 10.2 percentage points. This shows a clear polarization in the population percentages of Chinese cities of different sizes.

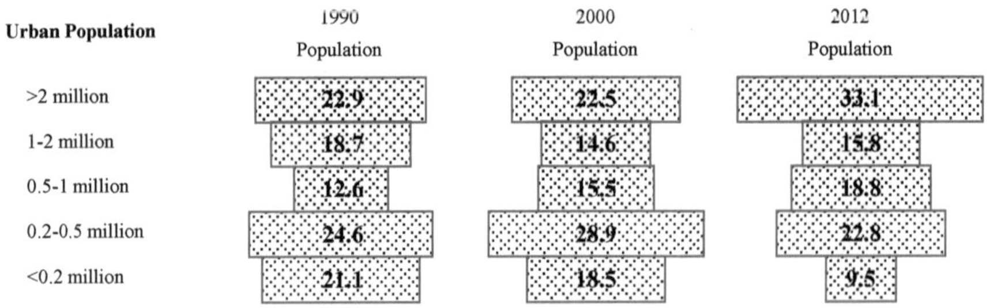

Fig. 4.2 Variation of China's urban population distribution by city size from dumbbell toward upside-down pyramid. *Note* Urban population sizes are non-agricultural Source: *China Population Statistics Yearbook* and *China Population & Employment Statistics Yearbook*

4.1.3 Polarization in the Ability to Accommodate Migrants by City Size

Since China's reform and opening up, great changes have occurred to the number and population percentages of Chinese cities of different sizes as well as their ability to accommodate migrants. To analyze changes in the ability of different sized cities to accommodate migrants, we collected statistics about changes in the percentages of total non-agricultural migrants in urban-area populations.

The growth rates of non-agricultural migrants to Chinese cities have generally changed in much the same pattern as changes to the population sizes of Chinese cities over the past 12 years (see Table 4.2). In this period, ultra-large cities (populations of over four million people) saw the highest population growth rate at 132.14%. Very large cities (populations of two to four million people) had a growth rate of 104.03%. In contrast, medium-sized cities had a population growth rate of only 18.52%, while small cities saw a decrease in their total population of 23.37%. With respect to the changes in the non-agricultural population percentages of cities of different sizes, the population percentages of ultra-large and very large cities rose by 7.13 and 3.46 percentage points respectively, whereas that of small and medium-sized cities dropped by as much as 15.07 percentage points. Interestingly, medium-sized cities saw their total non-agricultural population increase by 18.52%, but the overall population percentage still dropped by 6.02 percentage points. The aforementioned analysis shows that the populations of larger Chinese cities, especially very large and ultra-large ones, have been rapidly increasing in recent years, whereas the populations of small and medium-sized cities have been relatively shrinking, regardless of the number of non-agricultural migrants accepted or the non-agricultural population percentage.

The non-agricultural population of Chinese cities is much smaller than their permanent population due to the slow reform of China's *hukou* system. However, since statistics about the permanent population of China's urban system is not yet available, this study relies on urban area population data released by the Ministry of Housing and Urban-Rural Development (MOHURD) for analysis (see Table 4.3). In 2010,

Table 4.2 Changes in the numbers and population percentages of Chinese cities of different sizes

Urban population (in million people)	2000			2012			Changes in period 2000–2012			
	City number	Number share (%)	Population share (%)	City number	Number share (%)	Population share (%)	City number	Number share (in percentage points)	Population share (in percentage points)	Population growth (%)
>4.0	5	0.75	12.97	10	1.53	20.10	5	0.77	7.13	132.14
2.0–4.0	8	1.20	9.56	16	2.44	13.02	8	1.24	3.46	104.03
1.0–2.0	25	3.76	14.55	39	5.95	15.80	14	2.19	1.25	62.63
0.5–1.0	54	8.12	15.54	95	14.50	18.76	41	6.38	3.22	80.78
0.2–0.5	220	33.08	28.86	250	38.17	22.84	30	5.09	−6.02	18.52
<0.2	353	53.08	18.52	245	37.40	9.48	−108	−15.68	−9.05	−23.37
Total	665	100.00	100.00	655	100.00	100.00	−10	0	0	49.77

Note Urban population sizes are non-agricultural

Source: *China Population Statistics Yearbook* (2001) and *China Population & Employment Statistics Yearbook* (2013)

the total urban-area population of Chinese cities was 353.74 million people. Adding the temporary population of 40.95 million people, this total increases to 394.69 million people. This number is only 9.07 million people, or 2.2%, fewer than the urban permanent population recorded in the 2010 population census. Since the urban-area population of Chinese cities nearly equals their permanent population, it is feasible to categorize city sizes using statistics about the urban-area population. With respect to the number of migrant workers accepted by Chinese cities and thereby added to their urban-area populations, larger cities were the major recipients of migrant workers, receiving 83.95% of urban-area population additions. Among this total, ultra-large cities (with populations of greater than four million people) accepted 61.06%. In contrast, small cities saw a shrinking urban-area population due to the decreasing total number of small cities. With respect to urban-area population growth in this period, ultra-large cities saw their urban-area population grow by 23.60%, whereas small cities saw this population decrease by 17.34%. This is a typical example of polarization in China's urban system. Small and medium-sized cities are apparently unable to accommodate a greater number of migrants due to a decrease in the number of such cities and the suspension of city establishment procedures. Rather, new migrants have long been flocking to larger cities, especially ultra-large cities. This is a major reason behind the excessively fast expansion of some larger cities, and the growing crisis of urban ills.

Despite being large in number, Chinese towns tend to be small and financially weak, with an average town-area population of only 10,000 people (Wei 2010). Urbanization in China progressed very slowly between 1982 and 1990, when towns were fewer and had limited capacity to accommodate migrants. Nearly 75% of additions to the urban population were accepted by cities during this period (see Table 4.4). As Chinese towns rapidly grew in number between 1990 and 2000, they saw a sharp increase in the percentage of migrant worker additions to the urban population, reaching 49.94%, almost equal to the percentage of cities in this regard. Between 2000 and 2010, the percentage of Chinese towns in this regard dropped slightly to 47.39%, due to a roughly constant number of towns. Today, about 60% of China's urban population lives in cities and the remaining 40% lives in towns, according to the 2010 population census.

Based on the additions to urban-area population in the 2000–2010 period and 2006–2011 period, we can estimate that very large cities (with populations over one million people) have accepted about 36% of all additions to China's urban population in recent years. Large cities (with populations of 500,000–1,000,000 people) have accepted 8%, small and medium-sized cities (with populations of below 500,000 people) have accepted 9%, and towns have accepted 47%. The ratio among Chinese towns and cities with respect to the number of new migrants accepted is about 36:8:9:47. In other words, larger cities have accepted about 44% of all additions to China's urban population, whereas smaller cities and towns are far from able to accommodate the desired percentage number of new migrants, and will continue to decline relative to larger cities in their ability to accommodate migrants due to a decrease in city number. With their continuous influx of new migrants, larger cities, especially ultra-large cities, have seen excessively fast population increases

Table 4.3 Changes in the numbers and urban-area populations of Chinese cities of different sizes, 2006–2011

Urban population (in million people)	2006			2011			Changes in the period				
	City Number	Urban-area population (in million people)	Population share (%)	City number	Urban-area population (in million people)	Population share (%)	City number	Urban-area population increase (in million people)	Share of new migrants accepted (%)	Variation of population share (%)	Urban-area population growth (%)
>4.0	11	93.5867	25.11	13	115.6774	28.29	2	22.0907	61.06	3.18	23.60
2.0–4.0	22	59.0293	15.84	22	59.9705	14.67	0	0.9412	2.60	−1.17	1.59
1.0–2.0	35	49.2448	13.21	38	51.0631	12.49	3	1.8183	5.03	−0.72	3.69
0.5–1.0	92	63.8287	17.12	99	69.3487	16.96	7	5.5200	15.26	−0.17	8.65
0.2–0.5	230	71.509	19.19	267	83.4770	20.41	37	11.9680	33.08	1.23	16.74
<0.2	265	35.5289	9.53	218	29.3666	7.18	−47	−6.1623	−17.03	−2.35	−17.34
Total	655	372.7274	100.00	657	408.9033	100.00	2	36.1759	100.00	0.00	9.71

Note Urban-area population includes temporary population. Statistics about urban-area population in and before 2005 were not released
Source: *China Urban Construction Statistical Yearbook* (2006 and 2011)

Table 4.4 Acceptance of additions to China's urban population, 1982–2010

Indicator		City		Town		Total
		population (in million people)	Share (%)	Population (in million people)	Share (%)	
Urban population (in million people)	1982	145.2531	70.41	61.0561	29.59	206.3092
	1990	211.2225	71.32	84.9227	28.68	296.1452
	2000	292.6327	63.79	166.1383	36.21	458.7710
	2010	403.7600	60.26	266.2455	39.74	670.0055
Annual average urban population increase (in million people)	1982–1990	8.2462	73.43	2.9833	26.57	11.2295
	1990–2000	8.1410	50.06	8.1216	49.94	16.2626
	2000–2010	11.1127	52.61	10.0107	47.39	21.1235
	1982–2010	9.2324	55.75	7.3282	44.25	16.5606

Source Previous population censuses in China

and spatial expansion. This has caused these larger cities to quickly approach, and even exceed their overall carrying capacity. Additionally, unreasonable planning and ineffective management have caused serious problems such as high housing prices, traffic congestion, environmental degradation and growing social conflicts.

4.2 The Causes of Polarization from Multiple Perspectives

The polarization in China's urbanization to date is inconsistent with the Chinese government's basic guideline for urbanization. Comprehensive analysis from multiple perspectives reveals the reasons for such polarization. Generally, the root cause of the polarization is a combination of factors including traditional ideas of development, biased resources allocation, the market's polarization effect, farmers' willingness to migrate and regulatory ineffectiveness.

4.2.1 The Effects of Traditional Ideas of Development

Due to misconceptions about urbanization, local governments have tended to excessively emphasize size and GDP growth as their metrics for urbanization. Under this "traditional" mentality of development, excessive attention is paid to the economic benefits available in larger cities while the environmental impact, livability of cities, the costs of traffic congestion, and the skyrocketing living costs of urban residents are neglected. Under the current system, the benefits of the economies of scale that can be achieved in larger cities have been grossly overestimated, while the costs have been underestimated. Land and housing prices are high and rising fast in larger cities. While this is an important source of economic and wealth growth and has increased the output efficiency of larger cities, it drives higher prices of goods and living costs in larger cities, where commuting is time-consuming and costly. Unfortunately, however, scholars have failed to factor in rising costs of living when estimating the economies of scale of cities. This has doubtlessly led to overestimation. Larger cities consume enormous amounts of resources such as energy sources and water every day. These resources come from surrounding areas and elsewhere. With the large-scale movement of these resources, interregional value transfer will occur due to price distortion and an imperfect resources payment mechanism. As an example, the high costs of large water diversion projects implemented to alleviate water shortages in certain regions, especially larger cities, are not fully covered by water utility costs in large cities, but are shared across society and across regions. Places in Hebei Provinces such as Zhangjiakou City supply large amounts of water to Beijing every year, but receive unreasonably low payments for the supplied water resources. This suggests that, with value transfer resulting from price distortion, some larger cities have occupied economic benefits that would otherwise belong to their surrounding and other areas, through unequal exchange. Consequently, the economic benefits attributed to larger cities have been statistically overestimated. Under the traditional ideas of development, such overestimated economic benefits have become an excuse for some scholars and government departments to claim that resources should be allocated in favor of larger cities, without considering whether excessive expansion has reduced livability or surpassed the environmental carrying capacity. In addition to the pursuit of increasing in size, larger cities generally seek to become multi-faceted, that is, become a center in terms of politics, sciences, technology, culture, transport, information, finance and economy. An excess of functions and organizations will necessarily result in urban sprawl.

4.2.2 Dual Bias in Resource Allocation

Urban development in China is very administrative, as the Chinese government allocates resources apparently in favor of administrative centers and larger cities. Such dual bias in resource allocation and the resulting mutually enhancing effects is the primary cause for polarization in urbanization across China in recent years. Such dual bias also exists in some other developing countries. Government departments tend to concentrate an excess of resources in larger cities, especially national capitals, resulting in continuous size expansion of larger cities (Gugler 1982). Unlike in the other developing countries, towns and cities in China are categorized at different administrative levels. These levels include: municipalities directly under the Central Government, municipalities at the sub-provincial level, municipalities at the prefecture, sub-prefecture and county levels, and towns. The different levels of towns and cities vary from each other with respect to the levels of government officials, administration, resource allocation and authority for policy/law making. Towns and cities at lower levels must obey instructions from those at higher levels. Such a hierarchical urban management system facilitates coordination between towns and cities at the upper and lower levels, but is prone to dual polarization in resource allocation due to power inequality and a wide social capital gap between them. In other words, various resources are concentrated in administrative centers and larger cities at higher levels. Under the effects of such an administrative hierarchy on urban resource allocation, resources have long been concentrated in administrative centers, such as the capital city of Beijing, other municipalities directly under the control of the Central Government, cities specifically designated in the state plan, provincial capitals and municipalities at the prefecture level. The Central Government invests an excessive amount of resources in Beijing, the capital city, the other municipalities directly under the Central Government and cities specifically designated in the state plan. Provincial governments do the same thing with their respective provincial capitals. Such bias toward administrative centers has resulted in extremely unfair urban resource allocation. More growth opportunities and resources are given to administrative centers at higher levels, allowing the rapid agglomeration of people and industries, sufficient availability of jobs, excellent public infrastructure, and rapidly increasing city size. In 2012, there were 13 ultra-large Chinese cities with urban-area populations of more than four million people (including temporary population). These cities were Shanghai, Beijing, Chongqing, Shenzhen, Guangzhou, Tianjin, Wuhan, Dongguan, Zhengzhou, Shenyang, Nanjing, Chengdu and Harbin. All of these cities are either municipalities directly under control of the Central Government, cities specifically designated in the state plan or provincial capitals. The one exception is Dongguan, which would have an urban-area population of below two million people if its temporary population, which accounts for about 70% of its urban-area population of over six million people, were excluded. The capitals of all the 27 provinces and autonomous regions in the Chinese Mainland serve as their respective primary cities, with exception of Shandong Province and Fujian Province. In Shandong province, the city of Qingdao is slightly more populous than the provincial capital of Jinan. In Fujian, the city of Xiamen is slightly more populous than the provincial capital of Fuzhou. Such bias toward administrative centers is a de facto bias toward larger cities, the two of which are mutually enhancing. In addition, under the current system where counties are administered by cities, central cities at the prefecture level generally attract people, industries and resources from the administered counties through their administrative advantages. This has somewhat deprived the administered counties of growth opportunities, and the desirable ripple effects of these central cities have yet to occur. Today, it is a prevalent phenomenon among Chinese cities at the prefecture level for human, financial and land resources to be concentrated, or "shifted", into these cities from the administered counties.

4.2.3 The Polarization Effect of the Market

Cities are where people, the factors of production and non-agricultural industries agglomerate. In other words, they are important spaces where socioeconomic activities occur. The agglomeration of people, the factors of production and non-agricultural industries into towns and cities has led to economies of scale in multiple respects, such as the sharing of indivisible things, intermediate inputs and workforce and knowledge spillover (Wei 2006). This phenomenon is widely recognized by scholars. Henderson (2003) believes that the agglomerative effects of cities are based upon economies of scale and the laws of the market and thus are able to produce spillover effects. Such agglomerative effects of larger cities therefore come with self-enhancing effects, which will stimulate the flow of more people and resources into larger cities while making the expansion of smaller ones difficult, thus bringing about polarization in city size expansion. In the early and intermediate stages of economic development, in particular, economies of scale from the agglomerative effects of larger cities tend to be higher than those of smaller ones. Some studies show that larger cities with populations between one and four million people have the highest net economies of scale, equivalent to 17–19% of the city's GDP (Wang and Xia 1999; Au and Henderson 2006). Some other studies show that the correlation between the efficiency and size of Chinese cities at the prefecture or higher levels looks roughly like an upside-down U-shaped curve, whose vertex is somewhere between 3,520,000–9,320,000 people (Wang 2012). Scholars have yet to reach a consensus on optimal city size, but one thing is for sure—under a certain size limit, a city's efficiency increases together with its size. Accordingly, the spontaneous effects of market forces will produce a kind of polarization effect in order to maximize economies of scale from agglomeration. This will induce the agglomeration of various factors of production and resources into cities of larger sizes or at higher levels, leading to continuous population and spatial expansion of larger cities. Excessive expansion of larger cities usually occurs if the government lacks effective development plans and regulatory policies. With the size expansion of larger cities and continuous upgrading of smaller ones, the world's urban population has been concentrating into larger cities since the middle of the 20th century, showing very obvious population polarization during urbanization. Between 1950 and 2010, the global number of large cities with populations of over one million people grew in number from 75 to 449. The percentage of the world's urban population living in these cities increased from 23 to 38%. Statistical estimates indicate that the number of such cities will increase to 668 and their percentage in the world's urban population will rise to 47% (see Fig. 4.3). Developing countries lacking in effective planning and

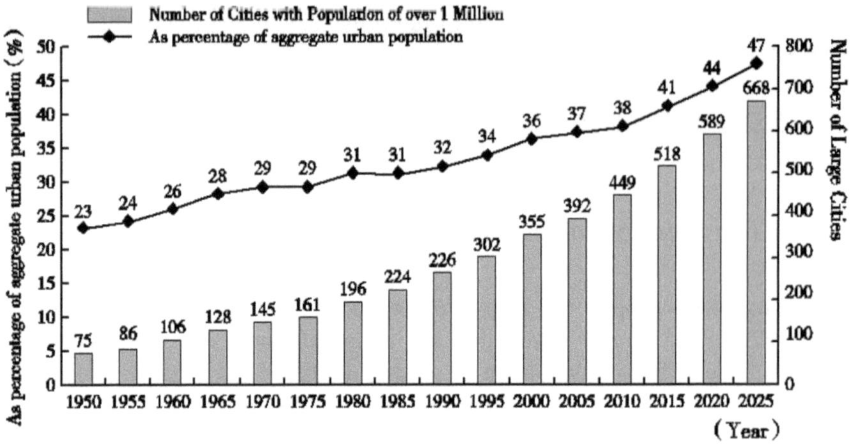

Fig. 4.3 Development of cities in the world with population over one million, 1950–2025 *Source* United Nations (2012)

regulation are the main contributors to this polarization in urban population. Some scholars refer to the phenomenon of a growing population percentage in large cities as "the ahead-of-the-times population growth law of large cities" (Hu et al. 1985; Hu 1986). Another scholar refers to it more simply as "the ahead-of-the-times *growth* law of large cities" (Gao 1991). Although the correctness of such a law remains in question, we are sure that, under the economies of scale from agglomeration, spontaneous market forces will produce a kind of polarization effect which stimulates the agglomeration of people, factors of production and industries into larger cities, worsening the developmental inequality between towns and cities of different sizes.

4.2.4 *Farmers' Willingness to Migrate*

The polarization effects of resource allocation by administrative means and by market forces have concentrated resources, factors of production and industries into larger cities as high-level administrative centers. Such concentration enables the governments of larger cities to provide better public services, conduct continuous improvements to infrastructure and the living environment, and increase residents' benefits, wages, and salaries. Consequently, there are huge gaps between towns and cities of different sizes in terms of employment opportunities, public services, wages/salaries/benefits and the living environment. When it comes to the quality of public infrastructure and the average amount of investment per capita in such infrastructures, at present Chinese cities fare higher than county seats, which in turn fare higher than general towns. In 2011, for example, the average amount of investment in public infrastructure per capita in Chinese cities was 3,461.43 yuan. This contrasts starkly with the 2,003.33 yuan investment per capita in county seats and 687.15 yuan investment per capita in general towns. In other words, Chinese cities were 1.73 and 5.04 times as high as county seats and general towns respectively. Large cities are also much higher than small and medium-sized cities when it comes to the quality of public infrastructure and services as well as average wage/salary per capita. In 2011, the thirty-six provincial capitals and cities specifically designated in the state plan were 15.51 and 58.43% higher than all cities at the prefecture or higher levels and cities at the county level respectively in terms of the average wage/salary per employee in city-administered areas. Excellent education, cultural and medical/health resources in China, in particular, are now highly concentrated in a small number of very large and ultra-large cities. Facing such huge gaps, farmers are more willing to migrate into large cities, especially very large and ultra-large ones, where there are sufficient employment opportunities, good public services and high incomes, rather than stay in towns and small cities nearer to their hometowns. This has hindered the progress of hometown-based urbanization and has also contributed to the dominant position of large cities in the urbanization process across China. In 2012, 65% of the 163 million non-local rural migrant workers in China lived in medium-size and large cities at the prefecture or higher levels, among which municipalities directly under the Central Government and provincial capitals accounted for 30.1% and cities at the prefecture level accounted for 34.9% (NBS 2013). Over 80% of migrants in China are located in medium-size and large cities, especially municipalities directly under the Central Government and cities specifically designated in the state plan and provincial capitals, which combined receive 54.1% of all migrants in China. About 70% of migrants who want to become registered city residents prefer larger cities (The Department of Services and Management for Migrant Population, the NHFPC 2012). In contrast, towns as well as small and medium-sized cities are much less attractive to farmers since they tend to lack industrial support, sufficient job availability and quality public infrastructures/services. This is in spite of the fact that the Central Government announced a few years ago that it would comprehensively lift registration restrictions to allow farmers to register as residents in towns and small cities, with similar systematic reform in medium-sized cities. With this in mind, it becomes clear that the main reason farmers prefer to move to larger cities versus smaller cities lies in the huge gap in terms of employment opportunities, wages/salaries/benefits, public services, and so forth.

4.2.5 Regulatory Ineffectiveness

In April 1997, the CPC Central Committee and the State Council released the *Notice on Further Enhancing Land Management to Effectively Protect Farmland*, which suspended the approval of applications for upgrading counties to cities in order to tighten control over the amount of land used for urban development, especially in small and medium-sized cities. Since this announcement, no new cities have been established with the exception of a few cases. Furthermore, some regions adjusted their means of administrative division, for example changing counties or cities into districts. As a result, the number of Chinese cities has actually decreased, rather than increased, declining from 668 in 1997 to 657 in 2012, despite a continuous increase in China's urban population and urbanization rate (see Fig. 4.4). In light of the declining number of cities caused by the indefinite suspension of city establishment, new migrants must be accommodated by existing cities, adding to the increasing expansion of city size.

Since the suspension of upgrading counties to cities, many cities at the prefecture or higher levels have been aggressively changing counties/cities into districts. This has resulted in a sharp increase in the number of city-administered areas from 727 in 1997 to 860 in 2012, whereas the number of cities at the county level decreased from 442 to 368. Due to the differences in administrative power between counties/cities and districts, changing the former into the latter brings the higher-level governments more benefits, including benefits related to land transfer. In addition to changing counties/cities into districts, medium and large city governments have also been busy setting up whole new districts. Both of these phenomena are fueling urban sprawl in larger cities. Under the current system, the ineffectiveness of urban planning and a lack of regulatory means make it difficult to effectively restrain the size expansion of very large cities, especially ultra-large cities with populations over four million people. A typical example in this regard is the city of Beijing, whose population limits specified in the city's overall plan have been frequently exceeded well ahead of time. According to the *Beijing Overall Urban Plan 2004–2020* approved by the State Council, Beijing's total population limit is supposed to be eighteen million

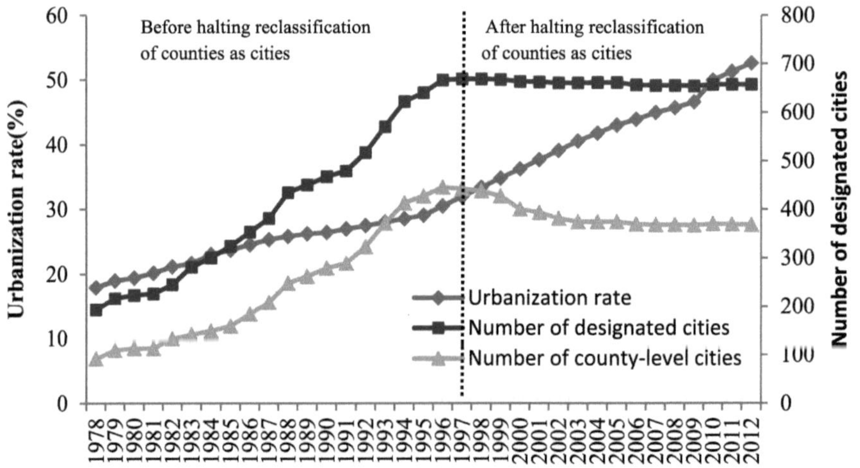

Fig. 4.4 Urbanization process and changes in the number of designated cities in China Source: *China Statistical Abstract* (2013) and *China Urban Construction Statistical Yearbook* (2012)

people by the end of 2020. In reality, however, that limit was exceeded in 2010. In 2012, Beijing had a permanent population of 20,693,000 people, including 7,738,000 people with a non-local *hukou*. Between 2001 and 2012, Beijing had an annual average net population increase of 588,100 people, including 431,400 people with a non-local *hukou*, in its permanent population. Driven by so-called "land-based revenues," Chinese cities, especially larger cities, are keen to sell land to increase local revenues. Some local governments use up all their land quotas for urban development, which were granted to the administered counties in central cities. The percentage of revenues from land transfer in the maintenance and construction expenditures of Chinese cities increased rapidly from 27.56% in 2006 to 58.78% in 2011.

4.3 The Scientific Basis for Resizing Towns and Cities

As a result of the polarization in urbanization across China, the size and structure of Chinese cities has become increasingly unreasonable, which in turn has hindered the effective, balanced allocation of spatial resources as well as the progress of urbanization and hukou urbanization. The polarization of city size and structure is detrimental to the creation of a scientifically reasonable urbanization pattern. Moreover, this polarization has induced a series of deep conflicts. First, there is the conflict between rapid population growth in very large cities and the carrying capacity of resources and the environment. With the large-scale agglomeration of resources, factors of production and industries, very large cities have more development opportunities, better job offers and higher capacity of accepting migrants. This has brought about a rapid increase in urban population. Between 2006 and 2012, twelve ultra-large cities with populations of at least four million people—including Shanghai, Beijing and Shenzhen—had an average increase in their urban-area population of 27.41%. This rate is much higher than the average increase of 13.29% among all Chinese cities. Unfortunately, however, these ultra-large cities generally face limitations with respect to their carrying capacity for resources and the environment. Some of these cities are approaching or have surpassed their carrying capacity. Today, mega-cities such as Beijing and Shanghai already suffer from population overload. In Shenzhen, for example, both chemical oxygen demand (COD) and NH3-N emissions are much higher than the local aquatic environment can tolerate (Wu et al. 2012). In recent years, Shenzhen in particular suffers unsustainability in terms of population, land, resources and the environment.

The second conflict is the conflict between the desire of towns and small/medium-sized cities to accelerate socioeconomic development and an insufficient capacity to receive migrants. Towns and small/medium-sized cities are large in number and widely distributed across China. They have lower entry barriers, sufficient carrying capacity of resources and the environment, and great growth potential. Accordingly, the Chinese government has long emphasized accelerated development of these towns and cities, a goal supported by the towns and cities themselves. However, towns and small/medium-sized cities are much less capable of accepting migrants, and are not sufficiently attractive to farmers given their lower-quality public infrastructure and services and lack of industrial support and employment opportunities. Third is the conflict between the eagerness of farmers to migrate into larger cities and the unaffordability of living in these cities. As mentioned above, farmers are most willing to work and live in larger cities given their better employment opportunities, public services and wages/benefits compared to small/medium sized cities. However, high housing and living costs in larger cities serve as high barriers to entry and hukou urbanization for farmers. It is very difficult for farmers to afford such costs with their low wages. It is estimated that large cities are two to three times as high, and very large/ultra-large cities three to five times as high as small and medium-sized cities when it comes to the average hukou urbanization cost per rural migrant worker [based on data of a research team from the former Ministry of Construction (MOC) 2006]. Non-local rural migrant workers across China have seen a faster increase in their monthly average income over the past few years, but this income is still much lower than the monthly average wage/salary of employees at organizations in their respective host cities.

As scientifically-based strategy to adjust China's city-size structure is required to address the aforementioned conflicts as soon as possible. Adjusting the overall city-size structure will allow the advantages of all towns and cities to be fully used, the consolidation of resources and division of labor to be more efficient, and the development of towns and cities of different sizes to be more fully coordinated in line with the goals of new urbanization with Chinese characteristics. With respect to the coordinated development of towns and cities of different sizes, the key is to rely on all towns and cities, rather than only larger cities and some small and medium-sized cities, to accept additions to the urban population. The city-size structure can be improved by optimizing the distribution of urban population. At present, the improvement of China's city-size structure relies on the consideration of several factors. The first factor is the carrying capacity of local resources and the environment. In particular, the carrying capacity of a locality's water resources, land and environment, should determine the maximum urban population that a city or region can afford. Approaching or surpassing this limit will place severe pressure on or cause severe damage to the environment, while reducing the city's livability and residential benefits. The carrying capacity of local resources and the environment therefore defines the upper limit of the urban population, namely, the maximum population that a city can afford under certain conditions. Measures such as extracting water from distant sources, utilizing resources in an efficient and effective manner, and promoting the improvement of the industrial structure can increase the carrying capacity of resources and the environment to a limited extent without reducing a city's livability and benefits available to its residents.

The second factor is the ability of towns and cities to accept migrants. This ability depends on the capacity of towns and cities to offer continuous industrial support and stable jobs. The ability of farmers to work and settle down in towns and cities requires stable jobs. Stable jobs in turn require continuous industrial support. If industries in towns and cities are not stable and subject to locational transfer as the environment changes, then jobs offered by such industries will be unstable, making it difficult for towns and cities to accept migrants continuously.

The third factor is the carrying capacity of urban public infrastructure. This includes the capacity of existing urban infrastructure in sectors such as public transport, medical/health care and education. From a short-term perspective, the capacity of this infrastructure will affect or even limit the population size that a town or city can afford. From a long-term perspective, however, this capacity will increase as investment in public infrastructure increases. The carrying capacity of urban public infrastructure is therefore changeable and will depend on future investment in such infrastructure. The fourth factor is the willingness of farmers to migrate. For towns and cities to accept rural migrants and to encourage farmers to relocate, full consideration of farmers' motivations and willingness to migrate is needed. Governments should follow the principles of voluntariness, categorization and adaptation to local conditions rather than force farmers to accept urbanization and hukou urbanization. Nowadays, Chinese farmers are less willing to become registered residents in towns and cities near to their hometowns, mainly because larger cities are more enticing in terms of job offers, development opportunities and public services. For this situation to change radically, the key is to narrow the gaps between larger cities and towns and small/medium cities in such aspects as infrastructure, public services and employment opportunities.

The fifth factor is the resumption of city establishment. China is and will continue for some time to be in a fast urbanization stage. With rapid urbanization and the constant growth of the urban population, it is unrealistic to rely only on existing cities to accept new urban residents, as this will further exacerbate the size issue of larger cities, especially very large and ultra-large cities. It is therefore necessary to resume the establishment of new cities as soon as possible. Counties and even towns that meet certain requirements should be gradually upgraded to city status at the county level in order to become an important force in receiving rural migrants.

After considering the aforementioned factors, we believe that the number of new migrants accepted by very large/ultra-large cities, large cities, small/medium-sized cities and towns should match a ratio of about 30:18:18:34 (see Fig. 4.5). The thinking behind this is as follows:

First, very large cities with populations of one to two million people still have great development potential, while cities with populations of two to four million people already face an increasingly limited carrying capacity with respect to resources and the environment, and ultra-large cities already suffer from population overload. Moreover, since an unreasonably large city size also increases a city's negative external effects, it is necessary to adopt measures to control urban populations (Wang 2010). As incomes continue to rise, urban residents will increasingly demand more recreational space and better living environments. Spaces only devoted to effective and efficient production will not be enough. However, maintaining quality living spaces and a beautiful environment will increasingly limit the capacity of larger cities with populations of one million or more people to accept new migrants. These cities will account for a lower percentage in the total number of new migrants accepted. Second, large Chinese cities with populations of five hundred thousand to one million

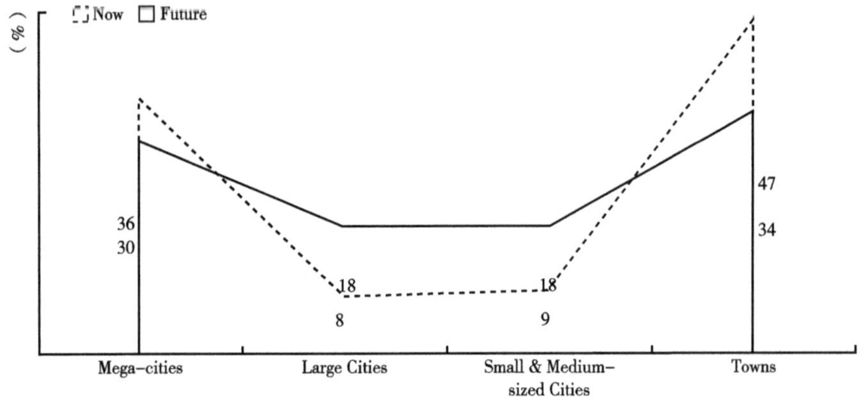

Fig. 4.5 Changes in the number of migrants absorbed by cities of different sizes as percentage of the national total

people will have greater agglomerative effects as a result of their current development stage. Moreover, some small and medium-sized cities will be upgraded to large cities, adding to the overall ability of large Cities to accept new migrants. Third, small and medium-sized cities will be able to accept a significantly higher number of new migrants than today thanks to continuous improvements in public services, stronger industrial support and the addition of new cities once China resumes the establishment of new cities. Lastly, Chinese towns will see a significant decrease in their acceptance share of the total number of new migrants once China resumes setting up new cities. This is mainly because some towns that meet certain requirements will be upgraded to cities.

Given the suspension of city establishment in China since 1997, many towns have approached or even surpassed the population threshold for setting up a city. According to the national standard for establishing cities approved by the State Council in 1993, a county whose population density per square kilometer is more than 400, 100–400 or fewer than 100 people may be upgraded to a city at the county level if there are no fewer than 120,000, 100,000 or 80,000 people who live in the county seat and work in non-agricultural sectors. The county must also meet other particular conditions. In special cases, a town that serves as the seat of the prefecture, league or county government may be upgraded to a city at the county level if it has a non-agricultural population of no fewer than 60,000 people. In 2011, 205 out of 19,600 towns across China had a population of over 100,000 people in their respective administered areas. More specifically, twenty such towns had a population of over 200,000 people while the remaining 185 towns had populations between 100,000 and 200,000 people (see Table 4.5). Based on their size, all of these towns could have been upgraded to small and medium-sized cities, with the exception of Humen Town, Dongguan City, Guangdong Province, whose population of over 500,000 people would have classified it as a large city.

Table 4.5 Population sizes of Chinese towns in 2011

Population size	Town number		Population in town-administered areas		Average population size (in people)
	Number	Share (%)	Population (in million people)	Share (%)	
>200,000	20	0.10	5.68	2.68	284,200
100,000–200,000	185	0.95	24.13	11.35	130,400
60,000–100,000	367	1.88	27.73	13.05	75,600
30,000–60,000	938	4.80	39.24	18.47	41,800
10,000–30,000	3497	17.88	58.07	27.33	16,600
5000–10,000	4386	22.43	30.92	14.55	7000
<5000	10,161	51.96	26.71	12.57	2600
Total	19,554	100.00	212.49	100.00	10,900

Source: *China Yearbook of Town Survey* (2012)

These 205 cities accounted for 14.03% of the total population of all Chinese towns and 4.32% of China's total urban population. Towns with populations of 60,000–100,000 people accounted for 4.01% of China's total urban population, and many of them also meet the conditions to be upgraded to city status. Moreover, some of these towns will inevitably become part of city-administered areas as medium-sized and larger cities continue to expand, subsuming towns into districts. As a result, many towns will become small and medium-sized cities or be included into city-administered areas if China resumes the establishment of new cities. This will significantly decrease the ability of towns to accept new migrants.

China is currently in a stage of strategic transition with respect to urbanization. The future focus of Chinese urbanization will center on promoting hukou urbanization and comprehensively improving the quality of urbanization. According to aforementioned forecasts, the urbanization rate in China will reach around 68% by the end of 2030 with a net urban population increase of 241 million people. By that time, large/ultra-large cities will have accepted 72 million new migrants, large cities 43 million, small/medium-sized cities 43 million, and towns 82 million new migrants. If this forecast proves accurate, the new migrant acceptance ratio among the different sized cities will be 30:18:18:34. In other words, larger cities with populations of at least 500,000 people will account for 48% of new migrants accepted in China, while towns and small/medium-sized cities will account for the remaining 52%.

4.4 A Strategic Choice for Resizing Towns and Cities

In summary, the main considerations for creating a scientifically sound city-size structure include adjusting to the carrying capacity of local resources and the environment, the capacity of urban public infrastructure, and the ability of cities to accept new migrants. This is a more expansive set of factors than the optimal city size argument advocated by some scholars. If an "optimal city size" existed, then all towns and cities would be of the same size and the varied system of towns and cities would not exist in the world. The optimal city size theory is obviously inconsistent with the real world. To scientifically guide city size restructuring, it is therefore necessary to enhance research on the motivating factors driving farmers to migrate to urban areas, as well as the overall carrying capacity of cities as soon as possible. City size regulation should be differentiated according to the carrying capacity of local resources and the environment, the capacity of urban public infrastructure and the ability to accept new migrants. Regional governments should reasonably direct the flow of rural migrants, rely on industrial agglomeration to guide the agglomeration of people, promote the co-agglomeration of industry and people, and organically combine industrial and urban development. They should also work to create a city-size structure characterized by urban agglomerations as the main form of urbanization, reasonable division of labor between towns and cities of all sizes, fully coordinated urban development and tiers. These steps will effectively roll back polarization in urban growth.

4.4.1 Establishing Scientifically-Sound City Size Standards

To address polarization in urbanization, some scholars and relevant government departments have proposed that city size standards be adjusted upwards. Namely, the standard for small cities should be increased from under 200,000 people to under 500,000 people; for medium-sized cities from 200,000–500,000 people to 500,000–1,000,000 people; for large cities from 500,000–1,000,000 people to 1,000,000–5,000,000 people; and for very large cities from over 1,000,000 people to over 5,000,000 people. Of course, other scholars have proposed other various standards for very large cities, such as an increase to 2,000,000 people (Yan 1989), 3,000,000 people (Niu et al. 2010) or even 10,000,000 people (Zhu 2013). Such proposals reflect the preference for larger cities and would cause more harm than benefit if they were incorporated into established state plan standards. Each of the proposed standards would statistically increase the number and population percentage of small and medium-sized cities and lead to a seemingly better city-size structure in China, but it would not eliminate the unbalanced city-size structure and moreover would produce various negative effects. International comparisons show that China's city size standards are already on the high side. The United Nations, for example, defines a city with a population of 100,000–1,000,000 people as a large city and a city with a population of over 1,000,000 people as a very large city. Globally, a city with a population of over 8,000,000 or 10,000,000 people is generally considered a mega-city. Such mega-cities can be both heaven and hell (Yang 2009), especially from such perspectives as livability, safety/security, harmony and sustainability. Both developed and developing countries alike target mega-cities as the main focus of population control and functional decentralization. More importantly, if China's city size standards were adjusted upwards, many medium-sized and larger Chinese cities would be downgraded, making it necessary for local governments to change relevant plans. Furthermore, downgrading would create a sense of loss in these cities, making them even more driven to expand in size. If the standard for very large cities increased to a minimum population of 5,000,000 people, China would have 11 very large cities based on urban-area population including temporary population. Besides the four municipalities directly under the Central Government, very large cities would only exist in the provinces of Guangdong, Hubei, Liaoning, Jiangsu and Henan. Other provinces and autonomous regions would certainly mobilize all possible resources to build very large cities of their own, resulting in yet another race for expanded city size. Increasing city size standards would therefore further intensify polarization in urban growth. With well-coordinated urban development in mind, we believe that the Chinese government should increase the number of levels for very large cities rather than simply increase city size standards. In other words, very large cities with populations over 1,000,000 people should be divided into three levels—1,000,000–4,000,000, 4,000,000–10,000,000 and more than 10,000,000 people. Cities at the second and third levels should be categorized as ultra-large cities and mega-cities respectively. Accordingly, Chinese cities would then be divided into six size levels: mega-cities (over 10,000,000 people), ultra-large cities (4,000,000–10,000,000 people), very large cities (1,000,000–4,000,000 people), large cities (500,000–1,000,000 people), medium-sized cities (200,000–500,000 people) and small cities (fewer than 200,000 people). For a large developing country like China, towns and cities will continuously grow in number during its stage of rapid urbanization, making it necessary to increase city size levels at the right time.

4.4.2 Differentiating Population Control in Very Large Cities

Very large Chinese cities and mega-cities are quickly approaching, or have already surpassed, the maximum carrying capacity of local resources and the environment. It is more necessary than ever to comprehensively control the population sizes of larger cities (with populations of more than four million people) as soon as possible given the limited environmental and resource carrying capacity, increasingly serious urban diseases, and negative external effects of agglomeration. The national central cities of Beijing, Shanghai, Tianjin, Chongqing and Guangzhou, and regional central cities of Shenzhen, Wuhan, Shenyang, Nanjing, Zhengzhou, Chengdu and Harbin are at the top tier of China's urban system and play a leading role in Chinese urban agglomerations. On one hand, it is necessary to maintain tight control over the population size in these cities in order to comprehensively tackle urban disease and prevent excessive spatial expansion. On the other hand, however, it is necessary to fully leverage these cities with respect to their leading role in economic development and ripple effects. The key to solving this dilemma is to change the pattern of development by promoting the comprehensive transformation and functional decentralization of cities. This requires replacing the current simple restrictions on migrants seeking to register as city residents with two-way, comprehensive regulation to control population sizes. On the one hand, it is necessary to determine the aforementioned cities' desirable functional position and developmental orientation, and then develop reasonable and viable industry entry standards based on these determinations. Emphasis should be placed on optimizing urban industrial structures to direct these cities toward high-end segments and services, thereby reducing the inflow of people through industrial optimization. On the other hand, it is necessary to direct the shift of people, industries, infrastructures and functions away from central urban areas in accordance with the carrying capacity of local resources and the environment and public infrastructure capacity. This guided shift will promote the optimization of the spatial structures and regional integration of larger cities, and increase their sustainability. Functional shifting, or functional decentralization, is critical for controlling the population sizes of larger cities. Therefore, regional governments should encourage people and industries to disperse from the central areas of larger cities into surrounding areas and provide guidance in this process in order to optimize the spatial structures of metropolitan areas. In addition, growth boundaries must be determined for larger cities, ecological red lines demarcated and the highest allowable development intensity determined based on overall carrying capacity, livability and sustainability, and balanced working, living and ecological spaces.

4.4.3 Unleashing the Potential of Urban Agglomerations as the Main Form of Urbanization

With the rapid emergence of urban agglomerations, China has entered a new stage of urbanization characterized by competition between urban agglomerations. Urbanization in China will continue at a rapid pace for the next two decades. Urban agglomerations are not only the main form of urbanization in China today, but they also play the leading role in accepting new migrants. Urban agglomerations will house over 60% of China's urban population by 2020, according to estimates from a research team at the Development Research Center of the State Council (2011). This means that urban agglomerations in China will become increasingly attractive to migrants in the future. Further unleashing the potential of urban agglomerations as the main form of urbanization will strengthen the leading role of these agglomerations in accepting new migrants. This study recommends the following steps to accomplish this.

First, develop urban agglomerations at different levels. From a long-term perspective, China should implement a multi-center network development strategy to build urban agglomerations of international, national or regional significance by 2030. These urban agglomerations will become an important engine for sustained, fast economic growth in China and serve as the leading carrier of China's urban population and non-agricultural industries.

Second, help urban agglomerations enhance sustainable development. For mature urban agglomerations such as the Yangtze and Pearl River Deltas and the Beijing-Tianjin-Hebei region, it is necessary to accelerate developmental transformation and industrial structure improvement with the limited carrying capacity of local resources and the environment in mind. In addition, these urban agglomerations should undertake measures to promote regional integration, optimize use of space, and enhance environmental governance and ecological restoration. These measures will enable these urban agglomerations to increase their international competitiveness, innovativeness and sustainability. For developing urban agglomerations in central and western regions, it is necessary to focus on enhancing the ability of these regions to accept a sufficient number of new migrants. It is also important to lower development levels, improve the investment environment, and build industrial support systems and functions of central cities. Enhancing industrial support in towns and cities is also important, as is improving division of labor and cooperation between towns and cities of various sizes. Relying on industrial support and environmental improvement will help unleash the growth potential of these towns and cities.

Third, promote the regional integration of urban agglomerations. Accelerating the building of rapid transport networks within urban agglomerations is a critical step to unleashing the potential of urban agglomerations. Priority should be placed on the development of public transport, especially rail transport. This should be accomplished with the goal of fueling integration with respect to infrastructure, industrial planning, environmental governance, markets for factors of productions, employment and social security. Measures should be taken to encourage rural migrants to live in the suburbs of central cities within urban agglomerations or in urban areas around these cities, and commute through the rapid transport system or work near their places of residence.

4.4.4 Improving the Ability of Town Sand Small/Medium-Sized Cities to Accept New Migrants

The overall carrying capacity of urban areas refers to the maximum population size that a town or city is able to house under conditions of a good environment and quality of life. A region's carrying capacity mainly depends upon the carrying capacity of local resources and the environment, the capacity of urban public infrastructure and the ability to accept new migrants. Today, larger Chinese cities are best able to accept large numbers of new migrants thanks to industrial agglomeration. However, they generally have a limited carrying capacity in terms of local resources and the environment. In contrast, towns and smaller cities usually have a higher environmental and resource carrying capacity, but have more limited capacity in terms of public infrastructure, industrial support, and the ability to accept new migrants. To address these issues, it is necessary to increase the environmental and resource carrying capacity of larger cities, on the one hand, and improve the ability of towns and smaller Cities to accept new migrants, on the other. As technology advances, the environmental and resource carrying capacity of larger cities can increase to a certain extent by increasing investment and adopting certain technologies. Nonetheless, excessive agglomeration of people and economic activities into larger cities will reduce urban livability and residential benefits, which contradicts the goal of people-oriented urbanization. From a livability perspective, it will be more feasible to improve the ability of towns and small/medium-sized cities to accept new migrants. Accordingly, local governments should focus on the main problems facing towns and small/medium-sized cities to address this issue, including improving public infrastructures and services, developing industries with local advantages and increasing employment opportunities. At the same time, resources must no longer be allocated in favor of administrative centers and larger cities. Instead, it is necessary to fully use the advantages of towns and cities of different sizes, improve functional allocation between towns and cities, and promote the comprehensive transformation of larger cities by shifting people, factors of production and industries away from central cities into towns and small/medium-sized cities. In addition, the Chinese government should offer incentives for private capital investment, both domestic and foreign, into towns and small/medium-sized cities, and favor these towns and cities when investing in urban infrastructure and public services. Financial subsidies, incentives and interest subsidies, in particular, should be offered to fuel the development of infrastructure, public services and affordable housing projects in towns. The Chinese government should encourage towns to develop in a manner characterized by specialization, high effectiveness and eco-friendliness. As an example, it should build specialized, livable towns with complete infrastructure networks, remarkable features, outstanding advantages and high competitiveness.

References

A Research Team at the Development Research Center of the State Council. (2011). *Civilianizing the rural migrants*. Beijing: China Development Press.

A Research Team of the Former Ministry of Construction (2006). *On new urban development requirements for rural migrant workers* (Report on Rural Migrant Workers in China). Beijing: China Yan Shi Press.

Au, C., & Henderson, J. V. (2006). Are Chinese cities too small? *The Review of Economic Studies, 73*(3), 549–576.

Gao, P. Y. (1991). *Comparative research on urbanization in China and other countries*. Tianjin: Nankai University Press.

Gugler, J. (1982). Over urbanization reconsidered. *Economic Development and Cultural Change, 31*(1), 173–189.

Henderson, J. V. (2003). The urbanization process and economic growth: the so-what question. *Journal of Economic Growth, 8*(1), 47–71.

Hu Z.L. (1986). On the ahead-of-the-times development of large cities and countermeasures. *Journal of Peking University (Philosophy & Social Sciences), 5*, 116–120.
Hu, Z. L., Liu, H. X., Ma, A. M., & Cai, Y. P. (1985). On the laws of the "ahead-of-the-times" population growth of large cities. *Social Survey and Research, 2,* 49–50.
National Bureau of Statistics. (2013). *Migrant workers monitoring survey report 2012*. May 27, 2013.
Niu, F. R., Bai, J. F., & Yang, Z. C. (2010). *Annual report on development of small and medium-sized cities in China (2010)*. Beijing: Social Sciences Academic Press (China).
The Department of Services and Management for Migrant Population, the National Health and Family Planning Commission of the People's Republic of China. (2012). *Report on China's migrant population development 2012*. Beijing: China Population Publishing House.
United Nations. (2012). *World urbanization prospects: The 2011 revision*. New York.
Wang, X. L. (2010). Economic analysis of the path to urbanization and city sizes in China. *Economic Research Journal, 10,* 20–32.
Wang, Y. Q. (2012). On the upside-down U-shaped city size vs. efficiency curve and its policy implications—Comparative research based on the economic, social and environmental efficiency of Chinese cities at the prefecture or higher levels. *Finance & Trade Economics, 11,* 127–136.
Wang, X. L., & Xia, X. L. (1999). Optimize city sizes to fuel economic growth. *Economic Research Journal, 9,* 22–29.
Wei, H. K. (2006). *Modern regional economics*. Beijing: Economic Management Press.
Wei, H. K. (2010). On the scientifically-sound development of town economies in China. *Jianghai Academic Journal, 2,* 80–86.
Wei, H. K. (2014). On polarization in urbanization and city resizing in China. *China Industrial Economics, 3,* 18–30.
Wu, B., Che, X. Z., Chen, X. D., & Xie, L. S. (2012). On Shenzhen's aquatic environment: Capacity and carrying capacity. *Research of Environmental Sciences, 25*(8), 953–958.
Yan, Z. M. (1989). On the criteria defining urban and rural populations and city size levels in China. *Population & Economics, 2,* 50–55.
Yang, B. J. (2009). Mega-cities: Both heaven and hell. *Business Watch Magazine, 1,* 62–63.
Zhou, Y. X. (2006). On urbanization rate in China. *Urban Planning, 30*(supplement), 32–36.
Zhou, Y. X. (2007). *Urban geography*. Beijing: The Commercial Press.
Zhu, H. (2013). Reinterpreting the concept of "very large city" in a new situation. *Regional Economic Review, 2,* 125–129.

Chapter 5
Optimizing the Spatial Distribution of Cities in China

City clustering has become a significant trend since China began reform and opening up. Both cities and urban residents are highly concentrated in coastal regions, and a number of urban agglomerations have developed. Cities and urban residents are generally concentrated along arteries. There are many reasons for this concentrated spatial distribution. Interregional differences in natural background conditions and resource endowment serve as primary reasons. China's development strategy in favor of eastern regions and the agglomerative effects of resource elements are also external conditions driving this trend. Yet the deeper, the basic driving force is the widening interregional gaps in economic development. The unbalanced spatial distribution of cities is creating an increasingly problematic situation of too high or too low urban population densities, lack of harmony between industrial and population distributions, heavy pressure on resources and the environment in certain areas, and insufficient interregional transport of energy sources and other commodities, and so forth. In order to improve the spatial distribution of Chinese towns and cities, the Chinese government should focus on better coordination of regional development in which not only the prosperity of people is emphasized, but also the prosperity of regions. For the future, the Chinese government should accelerate urbanization in central and western regions, implement a multi-center network development strategy, and build a system of major economic development belts led by the "four longitudinal and four latitudinal belts" concept. Measures should be taken to build a spatial distribution of towns and cities that match the carrying capacity of local resources and the environment.

H. Wei, *Urbanization in China*, Research Series on the Chinese Dream and China's Development Path, https://doi.org/10.1007/978-981-13-1408-7_5

5.1 Distribution of Chinese Towns and Cities: Agglomeration in Certain Areas

China's development strategy since reform and opening up has been implemented in a heavily biased manner. As a result of locational and policy advantages, eastern regions have been able to attract large amounts of overseas capital and technology. This has led to regional advantages in terms of socioeconomic development. The rapid agglomeration of industries in eastern regions has been driven by the interplay between factors of production such as people, money and technology. In addition, fast-growing eastern regions have had better success than central and western regions in terms of social development, residential income, public services and development opportunities. This has caused a rapid inflow of people from central and western regions into eastern regions. While the large-scale agglomeration of people and industries into eastern regions following China's reform and opening is consistent with the laws of the market economy and can be viewed as historically necessary, it has been executed in an uncoordinated and unbalanced manner. The agglomeration of people in eastern regions lags significantly behind industrial agglomeration. This has resulted in a serious lack of coordination in two respects. First is the lack of coordination between population and industry distributions. Eastern regions have sufficient jobs and opportunities but are short of available workers. In contrast, central and western regions have an ample workforce but are short of jobs and opportunities. Consequently, hundreds of millions of rural migrant workers move between eastern and central/western regions. The second is the lack of coordination between population/industry distributions and the carrying capacity of resources and the environment. The large-scale agglomeration of people and industry into eastern regions has already exceeded the capacity of local resources. As a result, eastern regions require the long-distance transport of large amounts of resources including energy sources from central and western regions. Such agglomeration has also posed a serious ecological threat to eastern regions. Meanwhile, central and western regions abound in land, minerals, energy sources and other resources, but are unable to attract and retain factors of production due to a lack of industrial support and employment opportunities. This lack of development coordination has caused serious problems including a widening socioeconomic development gap between regions, the interregional movement of hundreds of millions of rural migrant workers and commodities like energy sources, and heavy pressure on the environment and resources in certain areas. This chapter will present an in-depth discussion of the agglomeration of Chinese towns and cities in certain geographic areas. Such agglomeration manifests itself in three main respects as follows:

5.1.1 High Densities of Cities and Urban Residents in Coastal Regions

The high density of cities and urban residents in coastal regions is embodied by three major characteristics:

First, Chinese cities located to the southeast and northwest of the Aihui-Tengchong Line[1] show contrasting characteristics. The Aihui-Tengchong line is a geographical concept that divides China into southeastern and northwestern areas (see Table 5.1 and Fig. 5.1). After collecting data on the number and sizes of cities to both sides of this line, we find that the southeastern area accounts for only 45% of China's land area, but contains 89.65% of all Chinese cities (589 cities in total) and 94.3% of China's total urban-area population (385.606 million people). In comparison, the northwestern area accounts for 55% of China's land area, but contains only 68 cities, a mere 10.4% of all Chinese cities. It's urban-area population of 23.297 million people represents an astonishingly low 5.7% of China's total urban-area population. With respect to city size, of the 172 Chinese cities with an urban-area population of over 500,000 people, 163 (94.8%) are located in the southeastern area while only nine (5.2%) are located in the northwestern area. Among the 485 Chinese cities with an urban-area population of below 500,000 people, 426 (87.8%) are located in the southeastern area and 59 (12.2%) in the northwestern area. There are a total of 35 very large cities in China with an urban-area population of over two million people. 34 of these are located in the southeastern area. These statistics show that the spatial distribution of Chinese cities and urban populations is unevenly spread across China, and largely characterized by agglomeration. Not only does this reflect a serious imbalance in regional economic development, but this uneven spatial distribution places heavy strain on resources and the environment in certain areas due to the large-scale, interregional movement of commodities including energy sources.

The second major characteristic is that the number of cities in China's four main regions decreases sharply from east to west (see Table 5.2). According to data from 2012, East China accounts for only 10% of China's land area, but contains 35.6% of all Chinese cities (233 cities), and 48.8% of China's non-agricultural population. In contrast, West China accounts for 72% of China's land area, but contains only 25.2% of all Chinese cities (165 cities) 19.5% of China's non-agricultural population. With respect to the spatial distribution of cities based on size, East China accounts for 46.3% of all larger Chinese cities and, in particular, accounts for 55.4% of cities with populations of over one million people. In contrast, the much larger area of West China contains only16.3% of all larger cities in China, a total of 26. East China is home to 55 smaller cities, or 22.5% of all such cities in China, which is much lower than the region's percentage of larger cities. In comparison, West China is home to 80 smaller cities, or 32.7% of all such cities in China, which is much higher than its percentage of larger cities. In other words, the much smaller land area of East China has a much higher density of cities and urban population than West China. Moreover,

[1]Proposed by geographer Hu (1935), the Aihui-Tengcheng Line, also known as the Heihe-Tengchong Line, is a critical line that reflects the spatial distribution of China's population.

Table 5.1 City sizes in 2011: Southeastern versus Northwestern areas

Item		Total	>2.0 million	1.0–2.0 million	0.5–1.0 million	0.2–0.5 million	<0.2 million
China	Number of cities	657	35	38	99	267	218
	Population percentage (%)	100.0	43.0	12.5	17.0	20.4	7.2
Southeastern	Number of cities	589	34	33	96	245	181
	Population percentage (%)	94.3	42.2	10.6	16.5	18.7	6.2
Northwestern	Number of cities	68	1	5	3	22	37
	Population percentage (%)	5.7	0.7	1.8	0.5	1.7	1.0

Note Population refers to urban-area population, including temporary population
Source: *China Urban Construction Statistical Yearbook* (2011)

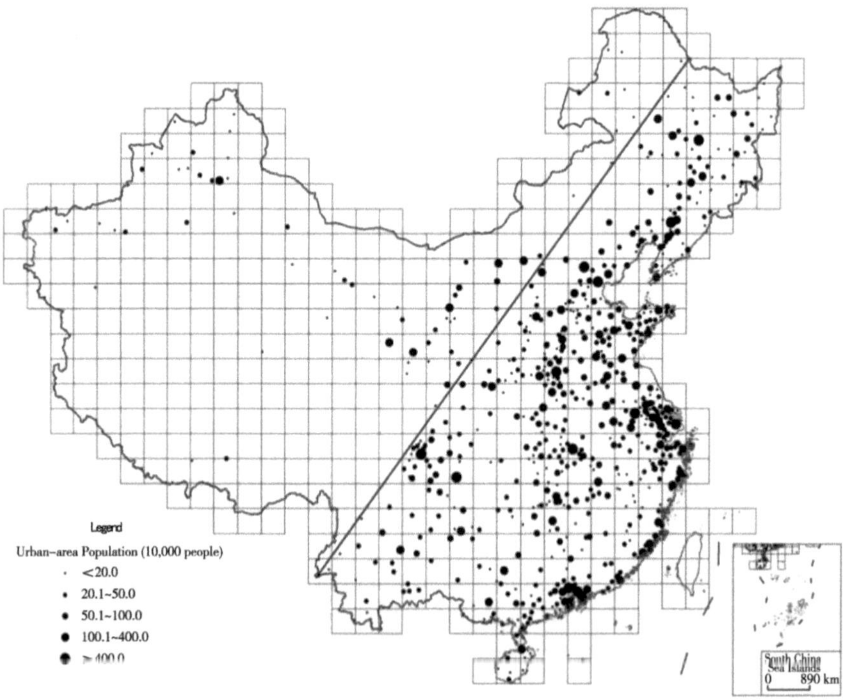

Fig. 5.1 City size in 2011: Southeastern versus Northwestern areas as divided by the Hu Huanyong Line. *Note* The straight line is the Aihui-Tengchong Line proposed by geographer Hu Huanyong in 1935. Source: *China Urban Construction Statistical Yearbook* (2011)

the larger the size of a city, the more likely it is located in East China. There is a clear and growing trend of Chinese cities and urban residents agglomerating into

East China. Although the number of cities in East China decreased from 249 in 2000 to 233 in 2012 due to the practice of changing some administered cities into districts, its percentage of China's non-agricultural population increased by 6.5 percentage points from 42.3 to 48.8%. During this same period, the number of cities in West China increased by seven, yet its percentage of China's non-agricultural population increased by a mere 0.4 percentage points. The number of cities in Central and Northeast China remained almost unchanged during this period, yet their percentages of China's non-agricultural population is in a clear downward trend, decreasing by 2.6 and 4.2 percentage points respectively between 2000 and 2012.

Third, the uneven distribution of Chinese cities is also embodied by inter-provincial differences in city number and density (see Fig. 5.2). From a city number perspective, there were 657 cities across China in 2011. Only 10 provinces had more than 30 cities, including Shandong, Guangdong, Henan, Jiangsu, Hubei, Hebei, Zhejiang, Sichuan, Liaoning and Heilongjiang. Combined, these provinces represent 55.3% of all Chinese cities. Of these, Shandong and Guangdong had the highest number of cities, Shandong with 48 and Guangdong with 44, or 7.3 and 6.7% of the total number of Chinese cities, respectively. On the other hand, there were four provinces and autonomous regions that had fewer than 10 cities, including Hainan, Ningxia, Qinghai and Tibet. The latter two had the smallest number of cities, two and three respectively. From a city density perspective, there were 0.078 cities per 1000 km^2 across China on average. The Jenks natural breaks classification method was used to group the city densities of Chinese provinces and autonomous regions into five classes. The highest city density class included the provinces of Jiangsu, Zhejiang, Shandong and Guangdong, with a city density of over 0.248 cities/1000 km^2. Jiangsu had the highest density at 0.377 cities/1000 km^2. The lowest city density class included Gansu, Inner Mongolia, Xinjiang, Qinghai and Tibet, with a city density of below 0.039 cities/1000 km^2. Tibet had the lowest density at 0.002 cities/1000 km^2. Generally, Chinese provinces with higher city densities are mostly in the economically developed eastern regions with good natural conditions. The provinces with lower city densities are mostly concentrated in the much larger but less developed western regions with poor natural conditions. Spatially, the number of cities decreases from east to central to west.

5.1.2 Urban Agglomerations

China has implemented T- and Π-shaped economic development strategies since it began reform and opening up. Urban agglomerations along arteries have emerged as a result of the distributions of important roads/railways and rivers as well as national strategic direction. We used a buffer and overlay analysis method (with a buffer distance of 50 km) to analyze how Chinese cities are spatially distributed along arteries (see Table 5.3 and Fig. 5.3). In relation to important rivers, urban agglomerations have appeared along the Yangtze, Yellow and Pearl Rivers. These three urban agglomerations account for 11.44% of China's land area, 19.33% of

Table 5.2 City sizes in China by region, 2012

Item		Total		>2.0 million		1.0–2.0 million		0.5–1.0 million		0.2–0.5 million		<0.2 million	
		2000	2012	2000	2012	2000	2012	2000	2012	2000	2012	2000	2012
China	Number of cities	665	655	13	26	25	39	54	95	220	250	353	245
	Population percentage (%)	100.0	100.0	22.5	33.1	14.6	15.8	15.5	18.8	28.9	22.8	18.5	9.5
East	Number of cities	249	233	5	14	10	22	18	38	95	104	121	55
	Population percentage (%)	42.3	48.8	12.2	20.2	5.9	8.9	5.4	7.7	12.4	9.9	6.5	2.1
Northeast	Number of cities	90	89	4	4	4	5	12	14	28	26	42	40
	Population percentage (%)	16.6	12.4	4.7	3.8	2.1	1.7	3.5	2.7	3.7	2.4	2.6	1.7
Central	Number of cities	168	168	1	4	6	6	19	27	50	61	92	70
	Population percentage (%)	21.9	19.3	1.9	3.6	3.6	2.5	5.0	5.2	6.6	5.4	4.8	2.7
West	Number of cities	158	165	3	4	5	6	5	16	47	59	98	80
	Population percentage (%)	19.1	19.5	3.7	5.6	2.9	2.6	1.6	3.2	6.2	5.1	4.7	2.9

Note Urban population sizes are non-agricultural
Source: *China Population and Employment Statistics Yearbook* (2013)

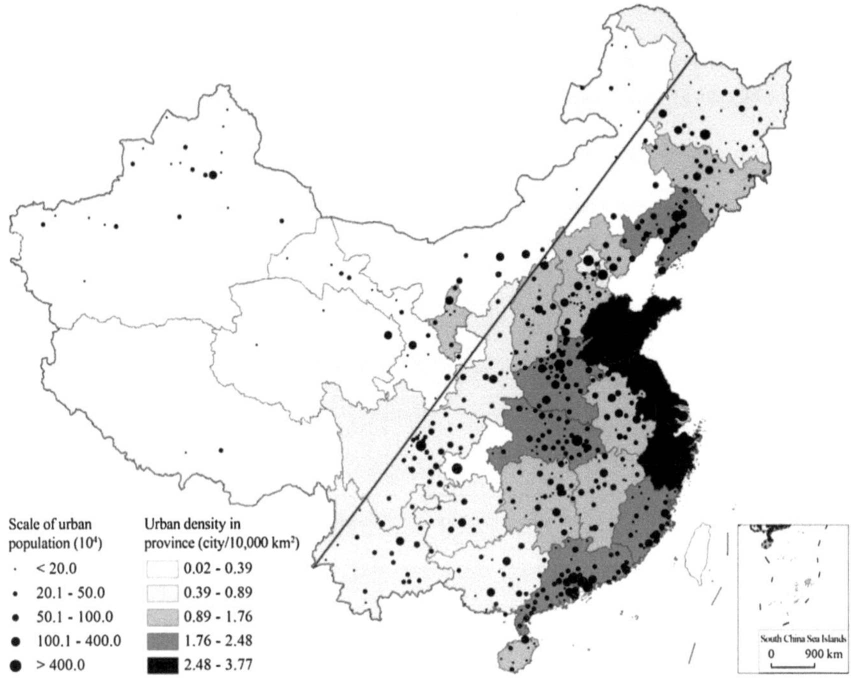

Fig. 5.2 City density in China. *Note* The straight line is the Aihui-Tengchong Line. Source: *China Urban Construction Statistical Yearbook* (2011)

all Chinese cities and 25.19% of China's urban population. These areas are 1.48 and 2.18 times higher than the national averages for city and urban population densities respectively. Of these urban agglomerations, the Yangtze River artery is the largest, housing 60 cities and over 47 million urban residents. In relation to important roads and railways, the primary urban agglomerations have appeared along the Beijing-Guangzhou, Beijing-Harbin, Beijing-Shanghai-Shenzhen and Lianyungang-Lanzhou-Urumqi lines. These four urban agglomerations account for 10.97% of China's land area, 34.86% of all Chinese cities and 55.71% of China's urban population. They are 2.78 and 5.03 times higher than the national averages for city and urban population densities respectively. They house 29 larger cities with populations over two million people, excluding the cities of Chengdu, Taiyuan, Dalian, Qingdao, Hefei and Nanning. Quality transportation infrastructure has enhanced inter-city connections and driven the economic development of cities along the aforementioned arteries. With the building of rapid transport networks, particularly the new high-speed railway network consisting of the "four longitudinal and four latitudinal lines",[2] China will continue to create a network of urban agglomerations.

[2]The four longitudinal lines refer to the Beijing-Shanghai, Beijing-Guangzhou-Shenzhen, Beijing-Harbin and southeastern, coastal Shanghai-Hangzhou-Fuzhou-Shenzhen passenger transport lines;

Table 5.3 Overview of urban agglomerations in China, 2011

Urban agglomeration	Area (km^2)	City number	Urban population (in million people)	City density (in cities/1000 km^2)	Urban population density (in people/km^2)
Yangtze River	479,252	60	47.850	0.125	100
Pearl River	183,733	23	27.471	0.125	150
Yellow River	435,714	44	27.684	0.101	64
Beijing-Guangzhou Line	228,607	68	75.777	0.297	331
Beijing-Harbin Line	168,570	32	48.148	0.190	286
Beijing-Shanghai-Shenzhen Line	314,664	98	140.065	0.311	445
Lianyungang-Lanzhou-Urumqi Line	397,697	48	31.292	0.121	79

Note Urban population refers to urban-area population, including temporary population
Source: *China Urban Construction Statistical Yearbook*

We adopted the concept of "space field energy" to describe the spatiotemporal distributions of Chinese cities, populations and the economy (see Fig. 5.4). From a regional development perspective, areas with high space field energy are strongly influenced by the region's central city, have fast and efficient connections within the region, and possess great growth potential. In contrast, areas with low space field energy have low ripple effects or are located far from the central regional city. Connections within the region are limited, and the opportunity for future growth is extremely limited (Guan et al. 2012). Figure 5.4 shows the number of urban agglomerations that have emerged in China. The Yangtze and Pearl River Deltas and the Beijing-Tianjin-Hebei region are relatively mature urban agglomerations with close intraregional economic connections. Urban agglomerations have also been growing in areas such as the central and southern Liaoning Province, Shandong Peninsula, Chengdu-Chongqing region, Guanzhong Plain, Henan Province, Wuhan-centric area and Changsha-Zhuzhou-Xiangtan area. As modern transport networks develop, some urban agglomerations have connected with each other. Along important rivers, the Yangtze River Delta, Hefei-centric, Middle Reaches of the Yangtze River and Chengdu Chongqing urban agglomerations have appeared. Along the Lianyungang-Lanzhou-Urumqi Rail Line, the Henan Province, Guanzhong-Tianshui, Lanzhou-Xining and Northern Slope of the Tianshan Mountains urban agglomerations have been established. Along China's coastline, there are the Central and South Liaoning, Beijing-Tianjin-Hebei, Shandong Peninsula, Yangtze River Delta, West Side of the Taiwan Strait, Pearl River Delta and Beibu Gulf (internationally known as the Gulf of Tonkin) urban agglomerations, which constitute China's most mature and impor-

the four latitudinal lines refer to the Xuzhou-Zhengzhou-Lanzhou, Shanghai-Kunming, Qingdao-Shijiazhuang-Taiyuan and Shanghai-Wuhan-Chongqing-Chengdu passenger transport lines.

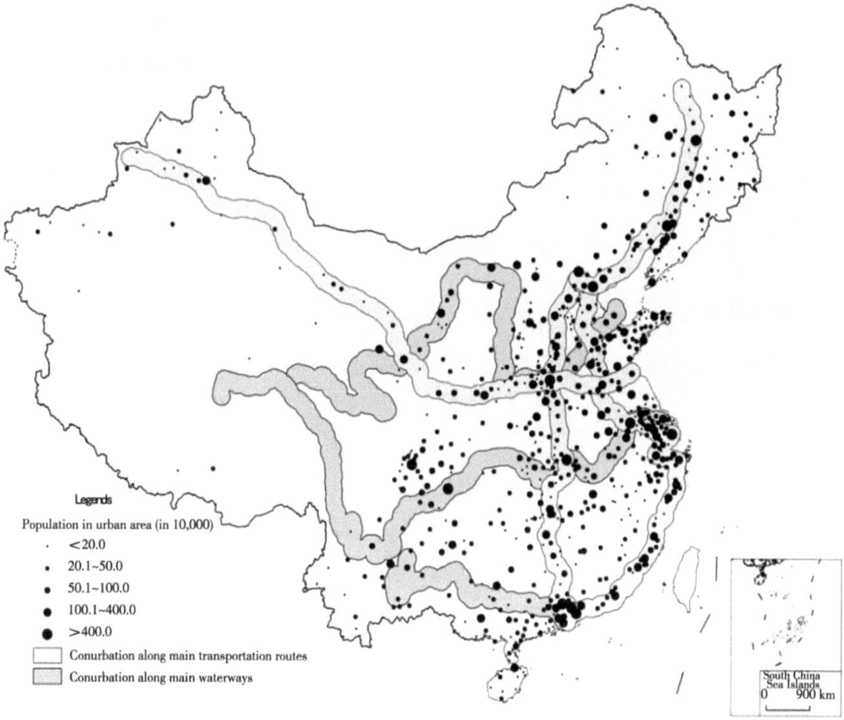

Fig. 5.3 Conurbation along arteries in China. Source: *China Urban Construction Statistical Yearbook* (2011)

tant population and industry clusters. Along the Beijing-Guangzhou Railway, there are the Beijing-Tianjin-Hebei, Central and South Hebei, Henan Province, Middle Reaches of the Yangtze River and Pearl River Delta urban agglomerations.

5.2 Main Problems Caused by Lack of Co-agglomeration

The agglomeration of Chinese towns and cities into certain areas is a result of a combination of factors including natural conditions, economies of scale, and historical factors. However, large-scale agglomeration of cities, people and industries into certain areas has not occurred without some challenges. Too high or too low urban population density levels are a growing concern. Unbalanced population and industry distributions have also created a widening interregional gap, interregional transfers of energy sources and other commodities, and unsustainable pressure on resources and the environment in certain areas. To a certain extent, this has threatened healthy, sustainable urbanization in China.

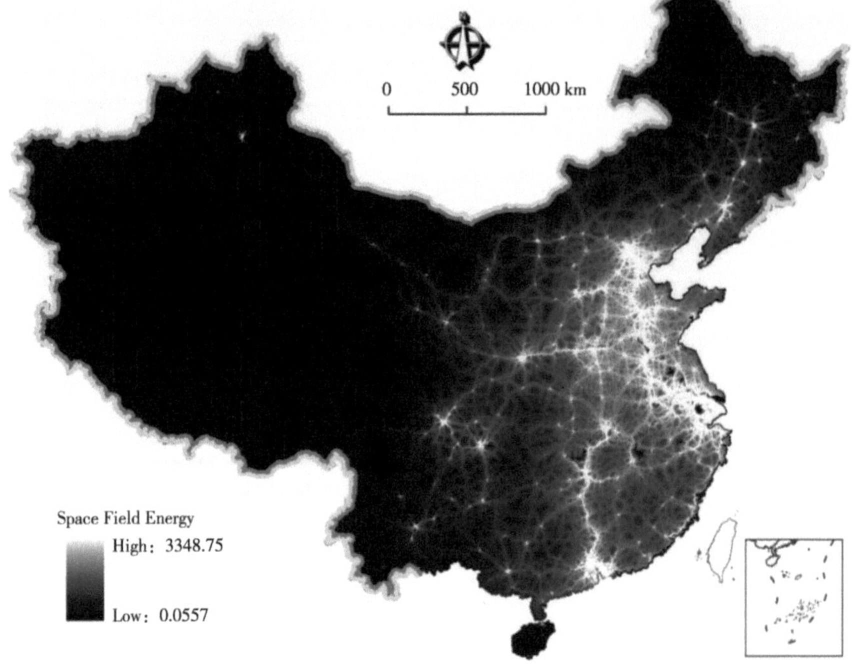

Fig. 5.4 Regional distribution of space field energy in China. *Source* The author himself

5.2.1 Too High or Low Urban Population Density

Too high or too low urban population density in China is manifested in inter-provincial differences in urban population density. In 2011, China had an urban-area population of 408.9033 million people, with an average population density of 43 people/km^2. The top five provinces by population density include Shanghai, Beijing, Tianjin, Guangdong and Jiangsu. These regions had population densities of 3761 people/km^2, 1062 people/km^2, 530 people/km^2, 267 people/km^2 and 265 people/km^2 respectively. The five provinces with the lowest population densities include Gansu, Inner Mongolia, Xinjiang, Qinghai and Tibet. These regions had population densities of 13.6 people/km^2, 7.3 people/km^2, 4.0 people/km^2, 1.8 people/km^2 and 0.4 people/km^2. Shanghai, the region with the highest population density, had a population density 9246 times higher than Tibet, the region with the lowest population density.

In order to conduct an in-depth analysis of interregional differences in urban population density, this study adopts a cell density analysis method that divides China into a grid of 150 km × 150 km cells, then counts the number of cities and urban population in each cell (see Fig. 5.5). The result shows that China's urban population is very unevenly distributed. Only 228 cells, or 44.4% of all cells, contain China's

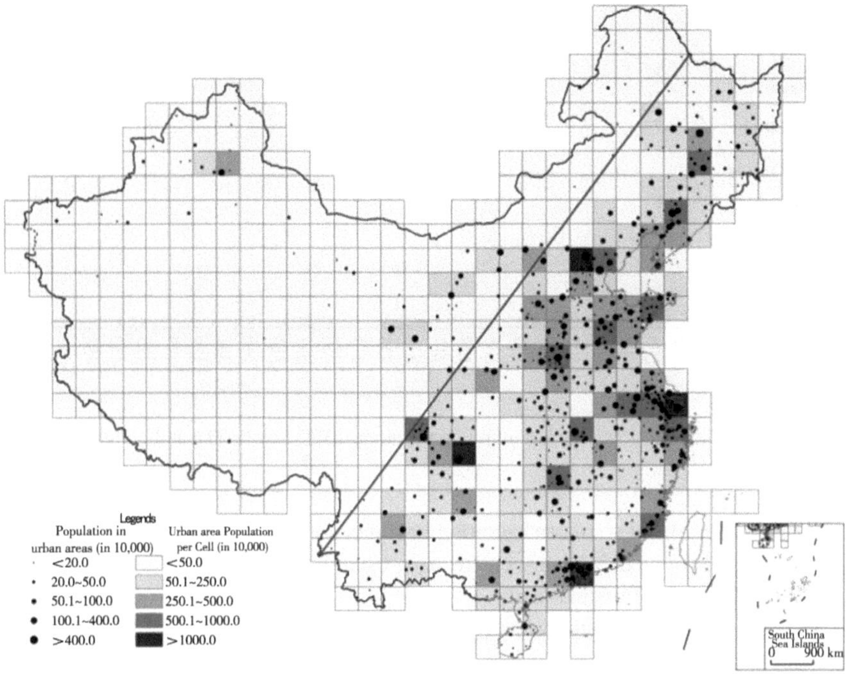

Fig. 5.5 Distribution of China's urban area population by cell density, 2011. Source: *China Urban Construction Statistical Yearbook* (2011)

entire urban-area population. The other 55.6% have no urban population at all. The most populous 22 cells, or 4.3% of all cells, house 207.85 million people, or over half of China's total urban population. According to the categorization results, there are four cells with urban-area populations of over 10 million people. These four cells have 31.31 million, 26.53 million, 18.18 million and 10.54 million people respectively, amounting to 86.56 million people, or 21.2% of China's urban population. In addition, there are 13 cells with urban-area populations of 5 to 10 million people. Combined, these 13 cells house an urban population of 93.51 million people, or 22.9% of China's total urban population. There are 29 cells with urban-area populations of 2.5 to 5 million people. These cells house an urban population of 102.08 million people, or 25% of China's total urban population. 88 cells have urban-area populations of 0.5 to 2.5 million people. These house an urban population of 103.25 million people, or 25.3% of China's total urban population. 94 cells have urban-area populations of below 0.5 million people. These house an urban population of 23.51 million people, or 5.7% of China's total urban population.

From a spatial distribution perspective, urban agglomerations with high population densities have emerged in the Yangtze and Pearl River Deltas as well as the Beijing-Tianjin-Hebei and Chengdu-Chongqing regions, followed by urban agglom-

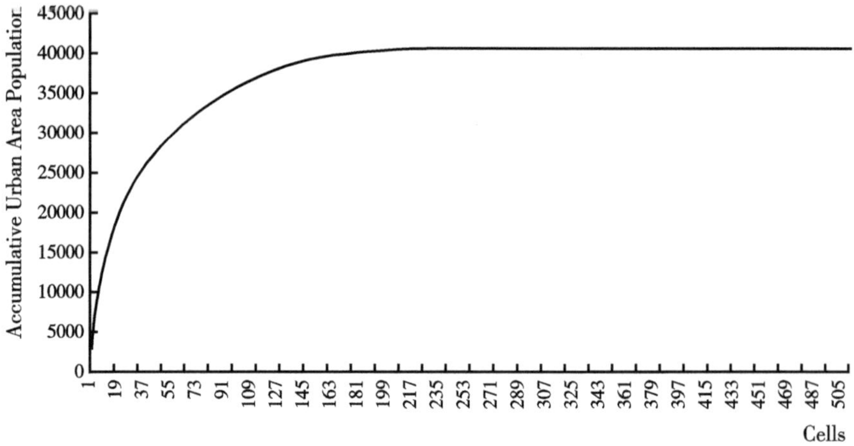

Fig. 5.6 China's urban area population in Lorenz curve, 2011. *Source* The author himself

erations in central and southern Liaoning Province, Shandong Peninsula, Guanzhong Plain, Henan Province, Wuhan-centric area, Changsha-Zhuzhou-Xiangtan area and the west side of the Taiwan Strait.

In order to reveal the characteristics of the spatial distribution of China's urban population, this study uses the Lorenz curve to analyze the urban-area population sizes of Chinese cities (see Fig. 5.6). The result also shows that China's urban population is very unevenly distributed. As mentioned above, only 228 cells, or 44.4% of all cells, house China's entire urban-area population, while the other 55.6% of cells have no urban population at all; the most populous 22 cells, or 4.3% of all cells, house 207.85 million people, or over half of China's total urban population. Excessively high or low urban population densities in China are creating unsustainable pressure on resources and the environment, and have even created safety risks in certain areas, particularly in some urban agglomerations with very high city and population densities.

5.2.2 *Extremely Unbalanced Regional Development*

Regional economic development has been extremely unbalanced since China began reform and opening up. Industrial agglomeration has occurred at a faster rate in developed eastern provinces than in central and western ones. At the same time, China has seen rapid changes in its population distribution as a result of the large-scale movement of its workforce. Nonetheless, there is a lack of coordination between people and industrial agglomeration; industries are highly concentrated in eastern regions with no corresponding human migration. This mismatch between the spatial distributions of population and industries has created extremely unbalanced regional

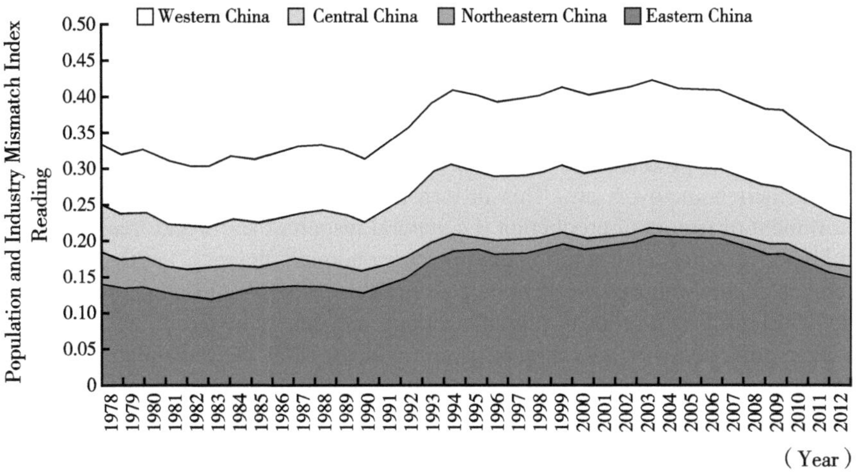

Fig. 5.7 Population and industry mismatch index in China by region. *Note* The M_i index refers to the absolute value of the difference between the percentage of the GRP of Region i in the sum of GRPs in the country and the percentage of its population in the national population. The equation is: $M = \sum_i M_i \sum_i \left| \frac{GRP_i}{\sum GRP_i} - \frac{POP_i}{\sum POP_i} \right|$. Source: *China Compendium of Statistics 1949–2004* and *China Statistics Yearbook* (2006–2013)

economic development in China (Cai and Zhang 2012; Wei et al. 2014). This study estimates the variation in population and industry mismatch, referred to as the M_i index (see Fig. 5.7). The national M index (as shown by the dotted line at the top) remained stable between 1978 and 1990, rose steeply between 1991 and 1994, stabilized again between 1995 and 2003, and dropped between 2003 and 2012. In this last period, the M index dropped from an all-time high of 0.4216 to 0.3219, falling very close to the level in the early stage of China's reform and opening up. Regionally, the M index of East China continuously rose from 0.1378 in 1978 to 0.2050 in 2003, before dropping to 0.1506 in 2012; the M index of Northeast China consistently dropped, while the M indexes in Central and West China remained stable. Interregional comparison shows that East China had the highest M index, higher than the indexes in West, Central and Northeast China by 1.66, 2.32 and 9.70 times respectively. This indicates that East China suffered from the largest mismatch between population and industry distributions. It accounted for 51.3% of the sum of gross regional products (GRPs) in China in 2012, but accounted for only 38.2% of China's total population. The imbalanced population and industrial co-agglomeration in East China has resulted in a wide interregional gap in socioeconomic development.

5.2.3 Interregional Transport of People and Commodities

The mismatch between population and industry distributions has resulted in a situation in which the location where products are manufactured is often different than where the products are sold, and where people work is often different from where their hometowns are. This in turn has caused the large-scale, continuous movement of factors of production (i.e. natural resources and workforce), resulting in higher materials consumption and lower economic efficiency. In 2013, the total number of rural migrant workers in China reached 268.94 million people, including 166.1 million non-local migrant workers. Of this latter figure, 88.71 million workers migrated within their respective provinces, while the remaining 77.39 million migrated between provinces. Non-local rural migrant workers from Central and West China, in particular, were most likely to migrate between provinces. Specifically, 62.5 and 54.1% of all rural migrant workers from Central and West China respectively had migrated between provinces. 66.02 million people, or 85.3% of all rural migrant workers who migrated between provinces, migrated to East China. Those who migrated to East China accounted for 89.9 and 82.7% of rural migrant workers from Central and West China respectively (NBS 2014). As a result of such large-scale movement of rural migrant workers, along with the considerable flows of students and people who travel for family reunions, the Spring Festival transportation season in China is believed to be the world's largest-scale human migration. The wide gap in socioeconomic development between the underdeveloped central and western regions and the developed eastern regions has contributed to this unusual phenomenon. In addition to this large-scale human migration, there is also an obvious mismatch between production and consumption of regional resources. In 2012, East China relied on neighboring Chinese regions for 79.0% of its coal, 94.7% of its oil products and 75.1% of its natural gas needs. In comparison, West China provided 57.1, 19.5, 95.5 and 54.6% of the total amounts of coal, oil products, natural gas and electricity respectively supplied to other provinces in China (see Table 5.4). It is apparent that China faces not only an uneven spatial distribution of energy sources but also a severe imbalance in the locations of energy production and consumption. China has embarked on a series of big projects to move resources, such as the north-to-south coal, west-to-east gas and electricity and south-to-north water projects. These initiatives have not only added to production costs, resource consumption and environmental degradation, but have also further widened the income gap between east and west as well as south and north due to the lack of corresponding compensation mechanisms.

Table 5.4 Energy consumption and inter-provincial energy allocation in China by region

Region	Coal consumption (in billion tons)			Oil product consumption (in billion tons)			Natural gas consumption (in billion m^3)			Electricity (in billion kWh)		
		From other provinces	To other provinces		From other provinces	To other provinces		From other provinces	To other provinces		From other provinces	To other provinces
East	1.540	1.217	0.173	0.265	0.251	0.215	6.209	4.664	0.204	2410.6	438.7	31.8
Central	1.092	0.389	0.675	0.075	0.064	0.010	1.917	1.582	0.156	1014.5	109.9	234.5
West	1.300	0.258	1.216	0.101	0.097	0.069	5.206	1.207	7.669	1282.0	76.8	374.3
Northeast	0.433	0.261	0.064	0.082	0.049	0.061	1.195	0.559	0	351.5	79.7	45.1
China	4.365	2.125	2.128	0.524	0.461	0.354	14.527	8.013	8.029	5058.7	705.1	685.6

Source: *China Energy Statistical Yearbook* (2013)

5.2.4 *Heavy Pressure on Resources and the Environment in Certain Areas*

The agglomeration of Chinese towns and cities has caused extreme pressure on local resources and the environment in certain areas. The consistent increase in economic activity and population density in urban and populous areas have imposed heavy pressures on resources and the environment, seriously threatening safety and regional sustainability. In recent years, urban-area population growth rates in ultra-large cities with populations over four million people, such as Shanghai, Beijing and Shenzhen, have been much higher than the average growth rate of all Chinese cities. Unfortunately, however, these ultra-large cities generally face limitations with regard to the local resource and environmental carrying capacity. In many cases these ultra-large cities have approached or even surpassed this capacity. According to a 2011 study, Beijing's population density rose from 766 people/km^2 in 1999 to 1230 people/km^2 in 2011.

Its permanent population of 20.186 million people in 2011 was already well beyond what the local land resources could support. The availability of freshwater resources is the greatest capacity concern in the Beijing-Tianjin-Hebei region. In 2011, the total amount of available water resources in Beijing was 2.681 billion m^3. Given the average water demand of 345 m^3/person, local water resources could support only 6.67 million people in Beijing, equivalent to 33% of its current population (Wen et al. 2013). The better-developed Pearl River Delta region also faces population pressures on the local resources and the environment. By the end of 2012, this region had a permanent population of 56.896 million people, with a population density of 1039 people/km^2. With 30.5% of Guangdong Province's land area, the Pearl River Delta accounts for 53.8% of its permanent population, 79.1% of its GRP and 83.6% of the gross industrial output value from enterprises at or above a certain revenue threshold. In the meantime, it consumes 73.6% of electricity, discharges 75.4% of wastewater and emits 61.6% of industrial waste gases in Guangdong Province. Today, the COD and discharged wastewater containing NH3-N in the city of Shenzhen are both far beyond what the local aquatic environment can support (Wu et al. 2012). Very large/ultra-large Chinese cities and the Pearl River Delta, Yangtze River Delta and Beijing-Tianjin-Hebei urban agglomerations all suffer from population overload. They will continue to face tough challenges in terms of population, land, other resources and the environment in the foreseeable future.

5.3 Analysis of the Lack of Co-agglomeration in Urbanization

The ongoing agglomeration of people and industries into eastern regions amid urbanization in China is a result of a combination of natural, policy, market and socioeconomic factors. China is a large country with significant interregional differences in natural background conditions and resource endowment. As a result, the emergence of uneven spatial distributions of China's population and economic activities is expected to a certain extent, and natural conditions are a large basis for the uneven distribution of Chinese towns and cities. However, there are several other factors at play. China has implemented a socioeconomic development strategy in favor of eastern regions since it began reform and opening up. With the majority of Central Government investment flowing into eastern cities, it is natural for large-scale industrial agglomeration to occur in eastern regions under basic market forces such as agglomeration effects and economies of scale. From a deeper perspective, it is the combination and interplay between natural conditions, a government development strategy that favors eastern regions, and market economy forces that has resulted in a widening gap in socioeconomic development between eastern and central/western regions. Such a gap is reflected in multiple dimensions including social development, local per capita income, public services, development opportunities, and so forth. This gap has contributed to the rapid flow of people from central and western regions into eastern regions. This widening interregional gap in socioeconomic development is now the primary reason for the rapid agglomeration of people and industries into eastern regions.

5.3.1 Interregional Differences in Natural Conditions

The world's third largest country in terms of land area, China has very different natural conditions among its regions. From a topographical perspective (see Fig. 5.8), China's land elevations decrease from west to east and can be roughly divided into three steps. The first step, represented by the Qinghai-Tibet Plateau, is on average more than 4500 m above sea level. The second step, which includes the Inner Mongolia, Loess and Yunnan-Guizhou Plateaus as well as the Junggar, Sichuan and Tarim Basins, is an average of 2000–3000 m above sea level. The third step covers areas less than 1000 m above sea level and mainly consists of plains and hills. As a whole, the first and second steps mostly encompass western regions with low city and population densities, whereas the third step mainly encompasses eastern, central and northeastern regions where the vast majority of Chinese cities and people reside. From an ecological perspective, China has a vast quantity of ecologically vulnerable areas. According to data from the "National Plan for Main Functional Areas," moderately vulnerable to extremely vulnerable areas account for 55% of China's land area. Of this percentage, 9.7% of China's land area is classified as extremely vulnerable, 19.8% severely vulnerable and 25.5% moderately vulnerable. From a development ability perspective, China is extremely mountainous, with mountains and plateaus accounting for about 60% of its land area. China's land area suitable for industrialization and urban development is about 1.80 million km^2. After excluding protected farmland and land already under use, China's land area still available for industrialization, urban development and other purposes is only about 280,000 km^2,

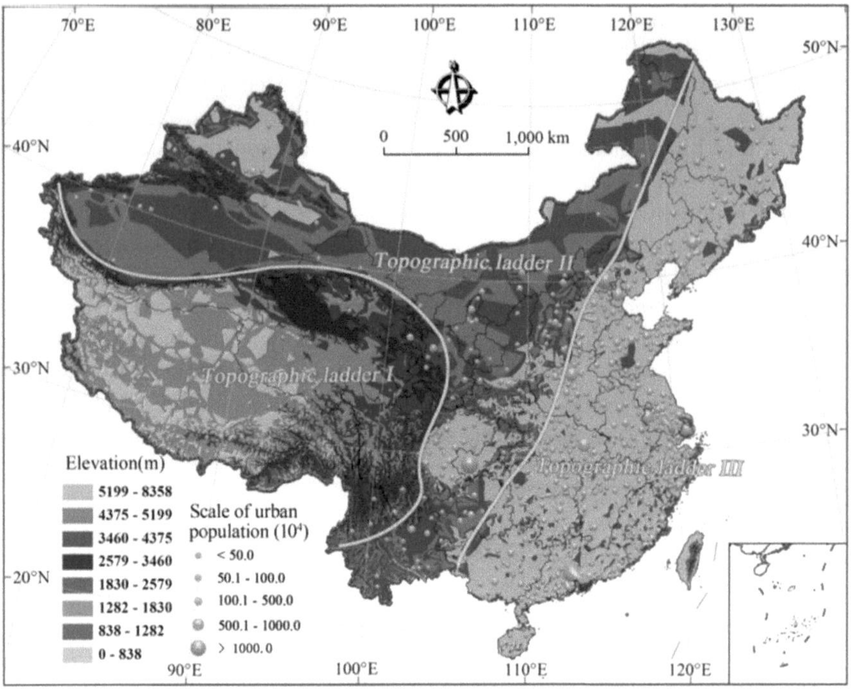

Fig. 5.8 China's triple level land elevation structure. *Source* The author

which accounts for only 3% of China's total land area.[3] Thus, despite China's large land area, very little land is available for development, much less urbanization and industrialization, as a result of interregional differences in land conditions, extremely vulnerable ecosystems in most regions and a relatively small land area suitable for development. Accordingly, Chinese urbanization, which involves over one billion people, must be conducted in a very limited area.

Interregional differences in natural conditions are a major contributor to the uneven spatial distribution of China's urban population. Back in 1935, geographer Hu Huanyong found out that population characteristics differed significantly on either side of the boundary between Aihui County, Heilongjiang Province and Tengchong County, Yunnan Province, subsequently known as the Hu Huanyong Line, which became the dividing line for the geographical distribution of China's population. The land-area ratio between regions southeast and northwest of the Hu Huanyong Line was 9:16. This is a stark contrast to the corresponding population ratio of 19:1 (Hu 1935). Radical socioeconomic changes have occurred in China since 1935, but the population distributions on both sides of the Hu Huanyong Line has remained generally unchanged. In 2011, for example, the urban-population ratio between regions

[3]Please refer to the *National Plan for Main Functional Areas—Creating an Efficient, Coordinated and Sustainable Pattern of Land Development*, December 2010.

southeast and northwest of the Hu Huanyong Line was 16.5:1, which was roughly consistent with the ratio in 1935.

5.3.2 *Resource Agglomeration Effect*

An agglomeration effect refers to the economic effects generated by the concentration of people, industries, resources and economic activities into a certain area. This effect is the primary reason for the emergence and expansion of cities. Urban areas are essentially spaces where people, money, technologies and industries agglomerate. The agglomeration of people and industry in turn shapes the appearance and evolution of urban areas. Industry agglomeration reduces costs, stimulates innovation, increases efficiency and encourages competition. An entire region can become more competitive and achieve external economic benefits through industrial agglomeration, and for this reason, industrial agglomeration generally leads to advanced industry clusters (Krugman 1991; Fujita et al. 2001; Wei 2003). The resource agglomeration effect is critical for the socioeconomic development of a region or even a whole country. When China relocated certain enterprises to inland areas, its goal was to even out the allocation of factors of production, but the result was reduced allocative efficiency of resources since it ran counter to the resource agglomeration effect. Since reform and opening, China has implemented a development strategy that prioritizes industry and cities. Such a pattern of socioeconomic development maximizes the resource agglomeration effect and economies of scale, thus leading to a successful outcome. In urban agglomerations, the spatial proximity of cities of varying sizes, human resources, technologies and enterprises is conducive to maximizing the agglomeration effect. This has led to cost and size advantages and allowed urban agglomerations to become the strategic growth engine of national and even global economies. For example, the urban agglomeration in the United States along the northeast Atlantic coast accounts for only 1.5% of the United States' land area, but accounts for 20% of the national population and 30% of the national manufacturing output. This urban agglomeration has long been the United States' largest production and commercial center and the world's largest financial center. The urban agglomeration in Japan along the Pacific coast contains over 60% of the Japanese population and accounts for 75% of its industrial output value. East China has grown rapidly since China began reform and opening up, with the most economically developed areas concentrated in the Pearl and Yangtze River Deltas as well as Beijing-Tianjin-Hebei. In order to achieve well-coordinated regional economic development and narrow interregional gaps, China is working to develop a number of urban agglomerations in central and western regions as well. This is consistent with the benefits of agglomeration effects.

5.3.3 A National Development Strategy in Favor of Cities and East China

China's development strategy has had a large impact on the agglomeration of Chinese towns and cities in East China in recent years. On one hand, biased socioeconomic policies in favor of cities over rural areas have led to a growing amount of economic resources directed into cities (Lu and Chen 2004). This has resulted in a widening income gap between urban and rural areas and a continuous flow of rural people into cities, especially larger cities. Accordingly, some cities have expanded rapidly and now suffer from unacceptably high population densities and environmental and resource problems. On the other hand, a national development strategy in favor of East China implemented since China began reform and opening up has led to the rapid growth of population and industries in eastern cities, as well as ever expanding sizes of larger cities. After the founding of the People's Republic of China, the Chinese government implemented a balanced regional development strategy in an attempt to narrow interregional gaps by evenly distributing productive forces. However, this strategy overlooked the principle of efficiency with respect to economic development and the regional distribution of productive forces. This strategy sacrificed return on investment (ROI) and resulted in slow economic development and heavy losses across China. With this lesson in mind, after reform and opening, China implemented an efficiency-oriented, uneven development strategy with East China as the focus of economic development. The Pearl and Yangtze River Deltas and the Bohai Economic Rim (BER), in particular, have successively become major regions for national development. A great deal of money, resources, and allotment of project and land quotas has been allocated in favor of East China. Additionally, the national government has established special economic zones, open coastal cities, economic and technological development areas, high-tech industry development areas and bonded harbor areas in Eastern China. These special zones and areas feature many incentives that have attracted a large amount of capital and technologies from outside China. Moreover, these zones have a full range of systems for division of labor and industrial cooperation and have a plentiful, low-cost workforce from central and western regions. While the agglomeration effect is beneficial in some respects, the massive agglomeration of China's urban population into eastern regions has also had the negative effect of increased pressure on local resources and the environment and the interregional transport of commodities. For this reason, the Chinese government has placed greater emphasis on well-coordinated regional development since the late 1990s. It has successively implemented several strategies to fuel the growth of less developed regions, such as the Western Development Strategy, Rise of the Central Regions Strategy and Northeast Revitalization Strategy. This practice is important for reducing the pressure on resources and the environment in eastern regions and achieving a moderately balanced spatial distribution of China's urban population.

5.3.4 The Widening Interregional Gap in Economic Development

The widening gap in economic development between China's eastern and central/western regions is the result of a combination of factors, including natural conditions, China's national development strategy, and market economy forces. The economic development gap is also the primary contributor to China's uneven spatial distribution of towns and cities. In 2010, the coastal urban agglomerations in the Yangtze and Pearl River Deltas and the Beijing-Tianjin-Hebei region combined accounted for only 3.63% of China's land area, but housed 16.57% of the national population and accounted for 33.78% of the total GDP. These regions also accounted for 32.5% of total consumer goods retail sales and 74.48% of China's total import/export value (see Table 5.5). In comparison, central and western regions as a whole accounted for only 39.3% of China's GDP.

The interregional gap in economic development is first manifested in an interregional gap in industrialization, which is an important reason for the former. The distribution of China's manufacturing is characterized by obvious concentration in certain areas and long-term imbalance between eastern and central/western regions. This imbalance is a result of natural topographical and transport conditions, national development strategies and the high agglomeration of foreign investment into coastal regions. It is the uneven manufacturing distribution among regions that has caused a widening interregional gap in China. Geographically, China's manufacturing is highly concentrated in eastern regions. In 2011, China's ten eastern provinces accounted for 60.7, 58.6 and 62.9% of China's manufacturing in terms of total output value, payable value-added tax and number of employees respectively. China's 18 central and western provinces accounted for only 31.0, 34.3 and 30.6% respectively. The eastern provinces of Jiangsu, Shandong, Guangdong, Zhejiang and Liaoning combined represent nearly half of China's manufacturing. Historically, eastern regions have seen a growing overall share of China's manufacturing, rising from 55.48% in 1990 to 60.73% in 2011, up 5.25 percentage points. The cities of Shanghai, Beijing and Tianjin have seen their respective shares decreasing more or less. Of these, Shanghai, one of China's most important industrial centers in the period of the planned economy, saw its share of China's manufacturing drop rapidly from the number two spot in 1990 to number nine in 2011. The other seven eastern provinces have all seen growing shares, especially those of Shandong, Guangdong and Jiangsu. There has been no significant change in the overall share of central and western regions, but Henan Province has seen its share rise rapidly by 1.77 percentage points. The overall share of northeastern provinces has dropped significantly from 12.41% in 1990 to 8.26% in 2011, down 4.15 percentage points. The provinces of Heilongjiang and Liaoning, for example, saw their shares drop by 1.95 and 1.74 percentage points respectively (see Fig. 5.9). Overall, China's manufacturing has become increasingly concentrated in eastern provinces since reform and opening. A typical center-periphery pattern has appeared when it comes to the distribution of

Table 5.5 Main indicators of three coastal urban agglomerations, 2010

		Population (in million people)	Area (in 1000 km²)	GRP (in billion yuan)	Total investment in fixed assets (in billion yuan)	Total consumer goods retail sales (in billion yuan)	Total import/export value (in billion USD)
Total	Yangtze River Delta	84.9075	109.9	7038.176	3341.468	2395.653	1037.860
	Pearl River Delta	56.1639	54.7	3767.326	1135.580	1261.324	751.303
	Beijing-Tianjin-Hebei	81.1179	183.8	3959.867	2329.478	1445.788	425.752
	Total	222.1893	348.4	14,765.369	6806.526	5102.765	2214.915
Share (%)	Yangtze River Delta	6.33	1.14	16.10	12.01	15.26	34.90
	Pearl River Delta	4.19	0.57	8.62	4.08	8.03	25.26
	Beijing-Tianjin-Hebei	6.05	1.91	9.06	8.38	9.21	14.32
	Total	16.57	3.63	33.78	24.47	32.50	74.48

Source: *China City Statistical Yearbook* (2011) and regional statistical yearbooks

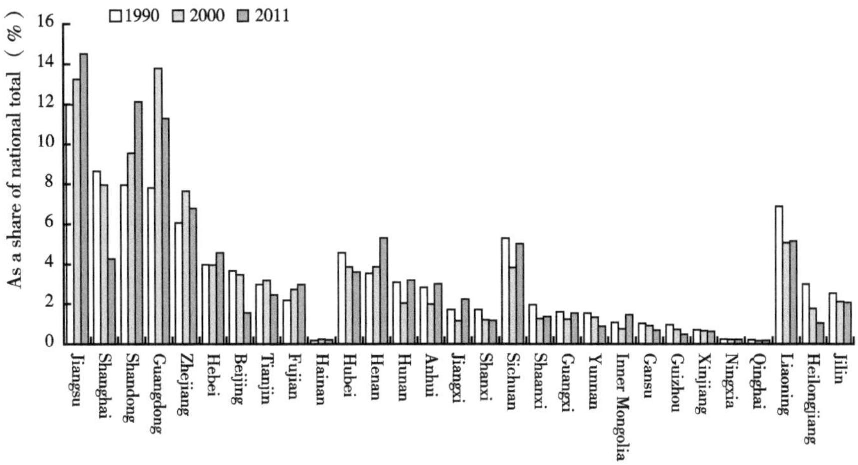

Fig. 5.9 Provincial manufacturing as a share of national total, 1990–2011. *Note* Only manufacturing was considered given that mining (B6–B11), wood and bamboo felling and transport (B12), electricity, steam and hot water generation and supply (D44), coal gas generation and supply (D45), tap water generation and supply (D46) and other industries rely on local natural resource endowment and market demand. Source: *China Industry Economy Statistical Yearbook* (1991, 2001 and 2012)

industries between the developed eastern and less developed central/western regions in China.

With respect to industry levels, there is an even greater interregional imbalance in the development of high-tech industries (see Table 5.6). In 2011, the total output value of China's high-tech industries was 8843.4 billion yuan. Regionally, eastern regions accounted for 78.84% of this total output value, 75.87% of the total number of employees in China's high-tech industries and nearly 80% of the R&D staff (full-time equivalent, or FTE). In contrast, the 18 central and western provinces combined accounted for only 17.43, 20.65 and 16.99% of the same indicators respectively. In particular, the high-tech industrial centers in the BER around Beijing and Tianjin, in the Pearl River Delta around Shenzhen and Guangzhou and in the Yangtze River Delta around Shanghai have become China's high-tech industry clusters.

5.4 Strategies to Improve the Distribution of Towns and Cities

Improving the spatial distribution of towns and cities requires coordinated regional development, narrowing the interregional gap in average resident income per capita and enabling access to the same basic public services and quality of life for all residents in all regions. In other words, it is necessary to emphasize not only the

Table 5.6 Regional changes in China's high-tech industries

Region	2000			2005			2012		
	Share of main business revenue (%)	Share of employees (%)	Share of R&D staff (FTE) (%)	Share of main business revenue (%)	Share of employees (%)	Share of R&D staff (FTE) (%)	Share of main business revenue (%)	Share of employees (%)	Share of R&D staff (FTE) (%)
East	80.4	60.8	49.6	88.5	78.0	65.0	76.6	73.9	79.6
Northeast	6.0	8.8	7.8	3.1	4.6	7.1	3.8	3.5	3.3
Central	6.4	13.6	15.2	4.3	8.6	13.4	10.9	13.2	9.7
West	7.2	16.8	27.5	4.1	8.9	14.6	8.8	9.3	7.4
Total	100	100	100	100	100	100	100	100	100

Source: *China Statistics Yearbook on High Technology Industry* (2012)

prosperity of people but the prosperity of regions during urbanization (Wei et al. 2011; Wei 2014). To achieve a balanced spatial distribution of Chinese towns and cities, the focus should be on addressing two issues—accelerating the westward industry shift and promoting eastward human migration. To this end, it is necessary to accelerate urbanization in central and western regions, keep tight control over the spatial expansion of towns and cities, implement a multi-center network development strategy, and establish a strategic geographical distribution of towns and cities characterized by the "four longitudinal and four latitudinal belts" concept. These efforts will lead to a more balanced spatial distribution of towns and cities.

5.4.1 Accelerating Urbanization in Central and Western Regions

Today, the urbanization rate in East China is already over 60%, meaning that this region is in the sub-stage of deceleration. From 2011 to 2012, the urbanization rate in East China grew an average of 1.08 percentage points per year, which was much lower than the growth of 1.43 percentage points between 2001 and 2005 and 1.59 percentage points between 2006 and 2010. Urbanization in East China is expected to slow down even further. Comparatively, the urbanization rates in central and western regions have yet to pass the inflection point of 50%. Urbanization in these regions will continue at a fast rate and may even accelerate in the near future. From 2011 to 2012, the urbanization rates in central and western regions grew an average of 1.81 and 1.66 percentage points respectively. These rates are much higher than in the 10th and 11th Five-Year Plan periods, as well as much higher than the growth rates in East China (see Table 1.3). Due to their different stages of urbanization—East China is in the deceleration sub-stage while central and western regions are in the acceleration sub-stage—the future focus of faster urbanization in China will be on central and western regions. Central and western regions will also play a leading role in accepting additions to the urban population during the future urbanization process in China. Accelerating urbanization in these regions will help local farmers achieve urban hukou urbanization near their hometowns and reduce the eastward flow of people so as to alleviate the pressure on the eastern urban agglomerations in the Pearl and Yangtze River Deltas and the Beijing-Tianjin-Hebei region. Moreover, it will help narrow the gap in economic development between central/western and eastern regions, thus promoting balanced spatial development and well-coordinated regional development.

For these reasons, the Chinese government announced at the Central Rural Work Conference measures to citizenize about 100 million people in central and western regions by the end of 2020. China's urban population is estimated to be about 126 million more in 2020 than in 2012 if its urbanization rate reaches about 60%. This means that nearly 80% of these new urban residents will be accepted by central and western regions before 2020. This is a rather difficult goal to achieve. From 2000

Table 5.7 Acceptance of new urban residents in China by region, 2000–2012

Region	Urban population (in million people)		Net increase	
	2000	2012	Number (in million people)	Share (%)
Total	465.67	716.25	250.58	100
East	203.54	318.33	114.79	45.8
Northeast	55.55	65.40	9.85	3.9
Central	104.49	169.55	65.06	26.0
West	102.08	162.97	60.89	24.3

Source: *China Statistics Yearbook* (2001 and 2013)

to 2012, China's urban population grew by 251 million people. Of this number, eastern regions accepted 45.8% of new urban residents, and central/western regions accepted 50.3% (see Table 5.7). There are two main drivers for achieving the goal of citizenizing about 100 million people in central and western regions by the end of 2020, assuming the urbanization rate in China tops the planned 60% target by the end of 2020. The first driver is accelerating urbanization in central and western regions; the second driver is increasing these regions' share of new urban residents accepted. In any case, however, it will be difficult to increase their current share of 50% to nearly 80%.

To accelerate urbanization in central and western regions, efforts should be undertaken in three main aspects. First, China should aggressively develop urban agglomerations. Urban agglomerations are already emerging in central and western regions, such as the Wuhan, Changsha-Zhuzhou-Xiangtan, Hefei, Henan Province, Chengdu-Chongqing and Guanzhong urban agglomerations. Careful planning and development of these urban agglomerations is needed to enable greater acceptance of new urban residents. Second, China should support the development of towns and smaller cities. A greater number of hometown-loving entrepreneurs and returned rural migrant workers are willing to invest and work in towns and smaller cities near their homes due to the advantages of lower living costs, lower entry barriers and ample human resources. For towns and smaller cities to attract investment, industries, projects and people, the key is to improve infrastructure, public services and relevant policies. Third, China should enhance industrial support in central and western towns and cities. Industries serve as the primary support and driving force for urbanization. Nowadays, China's manufacturing is highly concentrated in coastal regions, while its working population mostly comes from central and western regions. For people and industrial co-agglomeration to occur, the Chinese government must provide guidance and incentives for industries to shift from coastal regions into central and western regions, thereby creating more jobs. On the other hand, central and western regions should also undertake efforts to improve the environment and create conditions to incentivize the relocation of industries on a selective basis. They should organically combine industry incentives with local industrial structure improvement, and disallow the entry of heavily polluting industries.

5.4.2 Maintaining Tight Control Over the Spatial Expansion of Towns and Cities

The current high concentration of towns and cities in coastal regions, especially the Pearl River Delta, Yangtze River Delta and Beijing-Tianjin-Hebei urban agglomerations, has much to do with these regions' competitiveness and attractiveness in terms of employment opportunities, wages/salaries and public services. This concentration has also been intensified by arbitrary use of land and lack of consideration for the boundaries of ecological spaces and allowable land development intensity. For the future, it is necessary to maintain tight control over the spatial expansion of towns and cities, optimize the distribution of land used for urban development in each region, and define the allowable development intensity and boundaries of urban expansion, thereby preventing excessive and disorderly development in urban areas (Wei 2009a).

Optimize the allocation of land used for urban development. This is an important move for optimizing the spatial distribution of Chinese towns and cities. Nowadays, East China accounts for the majority of land used for urban development in China. In 2011, the ratio among East, Central and West China with respect to the area of land used for urban development was about 50.2:28.4:21.4.[4] For the future, this ratio should be adjusted to about 45:30:25 under the principle of "decentralization combined with concentration." With respect to decentralization, it is necessary to tighten control over land used for urban development in East China while gradually increasing land development quotas granted to Central and West China, which have relatively abundant land resources. This will lead to a better spatial distribution of land used for urban development throughout China. With respect to concentration, efforts should be undertaken to continuously increase the human carrying capacity and size efficiency of Chinese towns and cities so as to agglomerate more people. To improve spatial arrangements within urban areas, it is necessary to rely on advanced construction technologies and aggressively develop dense, high rise buildings and underground, multi-level infrastructure networks in central urban areas. Industrial parks should be built near each other, on the outskirts of urban areas according to national land use standards. The development of scattered, low-rise residential areas in suburbs should be tightly restricted, while dense, low buildings or sparse, high buildings should be encouraged (Liu 2014).

Establish Limits on Development Intensity. The recent influx of people and industries into East China, especially the Pearl River Delta, Yangtze River Delta and Beijing-Tianjin-Hebei urban agglomerations, has led to unacceptably high land development intensity in these regions. In contrast, Central and West China has

[4]Source: *China Urban Construction Statistical Yearbook* (2011).

abundant land resources, yet are still undeveloped (Wei 2009a). According to internationally accepted practices, the upper limit for land development intensity[5] in a country or region is 30%, beyond which the living environment for human beings is negatively affected. Land development intensity in developed countries and regions is typically under this limit. The land development intensity is only 15.6% in Japan's largest three urban agglomerations, 20% in Stuttgart, Germany, 21% in Paris, France, and only 19% in Hong Kong (Yang 2008). In contrast, the land development intensity in some Chinese cities is much higher: 47% in Shenzhen City, 43.8% in Dongguan City and over 40% in the entire Pearl River Delta (Zhang 2011). It is important that China strengthen space control and establish limits for the highest allowable development intensity. To optimize development zones, the overall development intensity should not exceed 25%. For cities within zones of national significance, the development intensity should not exceed 30%. For zones whose development is limited, the focus should be on point development with the highest level of development intensity set accordingly (Wei 2009a).

Define the Boundaries of Ecological Spaces. Nowadays, development has sprawled into surrounding rural areas and has continuously encroached on large amounts of farmland and ecological spaces. Some neighboring towns, cities and industrial areas have linked up into a single, vast "concrete jungle" with no ecological spaces. Some Chinese cities have even announced that they would let rural area disappear. Urban areas with insufficient ecological spaces are simply unlivable. The *National New Urbanization Plan (2014–2020)* has made it a goal for green space to account for 38.9% of developed urban land space across China by the end of 2020. From a large-scale perspective, however, preserving ecological spaces and other areas from development requires the establishment of boundaries in order to prevent urban spaces from sprawling indefinitely into a single whole. Boundaries for ecological spaces should be defined in strict accordance with the National Plan for Main Functional Areas (Wei 2009a), with different limits for different categories of development zone: development zones that require optimization, key development zones, zones with limited development, and zones with prohibited development.

Define Reasonable Percentages of Land Used for Different Purposes. The ratio of land used for different purposes within Chinese cities has become extremely skewed in recent years. The percentage of land space used for industrial production is relatively high, while that of living and ecological spaces is very low. In 2011, the total area of land used for urban development in China reached 41,805.25 km^2, of which land used for industrial and residential purposes accounted for 20.90% and 31.55% respectively, or a ratio of 1:1.5. Green spaces accounted for only 10.68% (see Table 5.8). In comparison, developed countries typically have 40–50% of land used for residential purposes in cities, and the typical ratio between land used for industrial and residential purposes is about 1:3. In France, the ratio is only 1:5, and in Japan, the ratio is an even lower 1:6. In China, not only is there a high percentage of land used for industrial purposes, but this land is also priced at obviously low levels.

[5]Land development intensity refers to the percentage of land used for development in the total land area in a region.

The average prices of such land in the Yangtze and Pearl River Deltas and the cities of Beijing and Tianjin are only 6.7–11.9%, 5.6–10% and 6.7–11.8% as high as the prices of industrial land in Japan's largest three urban agglomerations respectively.

The ratio between the average price of land used for industrial and residential purposes in China are also significantly lower than in developed countries such as Japan (Wei 2009a). In some Chinese towns and cities, land used for industrial purposes may actually be transferred at zero cost. In reality, local governments often push industrialization at the expense of land used for residential and ecological purposes. This practice is clearly inconsistent with the people-oriented Scientific Outlook on Development. To promote livability and attract business startups, sufficient space is needed not only for industrial production, but more importantly, for residential, ecological and recreational purposes. To this end, Chinese governments must follow the "people-oriented" principle in the process of urbanization by determining reasonable percentages of land used for living, ecological and production purposes, regulating land prices and setting the highest allowable percentage of land used for industrial purposes.

5.4.3 Implementing a Multi-center Network Development Strategy

Under the new circumstances of China's urbanization, the traditional uneven spatial distribution of towns and cities is increasingly at odds with the need for integrated development across China. It has even become a major obstacle to economic integration and well-coordinated regional development in this country. For the future, it is necessary to build a strategic framework of land development and urbanization in the form of a multi-center network as soon as possible. With urban agglomerations as central nodes and an integrated transport network as the connector, a multi-center network development strategy is needed so as to regulate spatial development and direct people and industrial co-agglomeration. This will promote the creation of a new pattern of development characterized by economic integration and coordinated regional development (Wei 2009b).

The so-called "multi-center approach" is geared toward achieving economic transformation in the Pearl and Yangtze River Deltas and the Beijing-Tianjin-Hebei region, while also developing a number of new growth engines in the BER and Northeast/Central/West China based on the development of urban agglomerations so as to achieve multi-center, diversified interregional competition. Since the 1980s, China's rapid economic growth has relied mainly on a few regions such as the Pearl and Yangtze River Deltas. Such a pattern of economic growth will no longer exist, according to relevant trends. Given the differences in development conditions and stages among urban agglomerations, the Chinese government should rely on major development belts to build a three-tier system of urban agglomerations throughout China. This will help shape these new development belts into the new leaders and

Table 5.8 Overview of Land used for development in Chinese cities of varying sizes, 2011

Urban-area population (in million people)	City number	Percentage of land used for urban development (%)								
		Residential	Public infrastructure	Industrial	Warehousing	Transport	Road and square	Municipal facility	Green space	Special
>2.0	35	29.82	12.39	21.23	4.33	4.67	12.18	3.63	9.75	2.00
1.0–2.0	38	30.14	11.76	23.88	3.41	4.69	10.99	3.27	10.22	1.63
0.5–1.0	99	32.58	12.09	21.29	3.36	4.22	11.08	3.17	10.82	1.39
0.2–0.5	267	33.16	11.92	19.19	3.62	4.53	10.61	3.89	11.58	1.49
<0.2	218	33.90	11.43	18.28	3.83	3.66	11.27	3.72	12.28	1.65
Total	657	31.55	12.04	20.90	3.79	4.46	11.34	3.56	10.68	1.68

Note Urban population sizes are grouped by urban-area population, which includes temporary population
Source: *China Urban Construction Statistical Yearbook* (2011)

Table 5.9 Three-tier system of urban agglomerations in China

Tier	Count	Urban agglomerations
World-class	4	In The Yangtze River Deltas region, the Pearl River Deltas region, Beijing-Tianjin-Hebei region and Middle Reaches of the Yangtze River
National	10	Chengdu-Chongqing region, Shandong Peninsula region, Guanzhong-Tianshui region, West Side of the Taiwan Strait, central and south Liaoning Province, Harbin-Changchun region, Lanzhou-Xining region, Beibu Gulf, central Henan Province and Hefei-centric region
Regional	9	Central and south Hebei Province, Taiyuan region, Hohhot-Baotou-Ordos region, central Guizhou Province, central Yunnan Province, Ningxia section of the Yellow River, the east end of the Lianyungang-Lanzhou Railway, the northern slope of the Tianshan Mountains and central and south Tibet

Source The author

growth engines for future rapid economic growth in China, and facilitate multi-center, diversified interregional competition (see Table 5.9). Tier 1 development belts would consist of the four world-class urban agglomerations in the Yangtze and Pearl River Deltas, Beijing-Tianjin-Hebei region and Middle Reaches of the Yangtze River. Tier 2 would consist of the ten urban agglomerations of national significance in the Chengdu-Chongqing region, Shandong Peninsula, Guanzhong-Tianshui region, West Side of the Taiwan Strait, central and south Liaoning Province, Harbin-Changchun region, Lanzhou-Xining region, Beibu Gulf, Henan Province and Hefei-centric region. Tier 3 would consist of nine regional urban agglomerations with influence only within their respective provinces, including central and south Hebei Province, Taiyuan region, Hohhot-Baotou-Ordos region, central Guizhou Province, central Yunnan Province, Ningxia section of the Yellow River, the east end of the Lianyungang-Lanzhou Railway, the northern slope of the Tianshan Mountains and central and south Tibet. Of these, urban agglomerations in the Yangtze and Pearl River Deltas and Beijing-Tianjin-Hebei region will likely become world-class urban agglomerations by 2020; those in the Middle Reaches of the Yangtze River by 2030.

This so-called "network development" will further improve the coastal development belt and enhance the development of the Yangtze River belt—especially its middle and upper reaches. Relying on important roads/railways and facilitating a larger integrated transport network will help accelerate the balanced development of urban agglomerations and central cities and establish new major development belts of national significance. The overall goal is to gradually establish a strategic, network-like pattern of land development and urbanization (see Fig. 5.10). By building a multi-center network, the Chinese government will be able to gradually develop a number of new growth engines for rapid economic development in China and promote the creation of an integrated, balanced spatial distribution of towns and cities (Wei 2009b).

5.4.4 Building a System of Major Development Belts Characterized by Four Longitudinal and Four Latitudinal Belts

In the *Opinions of the CPC Central Committee on Making the 10th Five-Year Plan for National Economic and Social Development* passed at the 5th Plenary Session of the 15th CPC Central Committee in October 2000, the Chinese government announced for the first time that it would duly implement an urbanization strategy and gradually build a reasonable urban system. Later on, in the *Outline of the National Urban System Plan (2005–2020),* the Chinese government announced that it would create a multi-center spatial distribution of towns and cities and build seven development belts intended to enhance interregional cooperation, including:

(1) The coastal urban belt of Beijing-Guangzhou (containing the Beijing-Kowloon Railway)
(2) Beijing-Hohhot-Baotou-Yinchuan-Lanzhou
(3) Xining-Chengdu-Kunming-Beibu Gulf
(4) Harbin-Dalian
(5) Yangtze River
(6) Lianyungang-Lanzhou-Urumqi
(7) Shanghai-Nanchang-Changsha-Guiyang-Kunming.

In the 11th Five-Year Plan released in March 2006, the Chinese government announced that it would gradually create a spatial distribution of towns and cities led by the coastal, Harbin-Beijing-Guangzhou, Yangtze River and Lianyungang-Lanzhou development belts as well as a number of urban agglomerations. In December 2010, it announced in the National Plan for Main Functional Areas that it would build a strategic distribution of towns and cities mainly covering the latitudinal Lianyungang-Lanzhou-Urumqi and Yangtze River development belts, the longitudinal coastal, Harbin-Beijing-Guangzhou and Baotou-Kunming development belts, focusing on development zones to be optimized, key development zones and other urban areas within these belts. It further clarified at the Central Urbanization Work Conference in December 2013 that it would make all possible efforts to achieve the aforementioned strategic distribution of towns and cities.

All the aforementioned national plans place great emphasis on the strategic distribution of towns and cities and have proposed corresponding strategic visions. Although released at different times, these plans all cover a period that ends in 2020, with the proposed strategic visions of urbanization focused mostly on short- or medium-term rather than long-term. In addition, the urban agglomerations proposed in the National Plan for Main Functional Areas vary widely in terms of coverage—the largest BER urbanized region includes the Beijing-Tianjin-Hebei, Central and South Liaoning and Shandong Peninsula urban agglomerations, whereas the smallest Central and South Tibet urbanized region only includes Lhasa and its surrounding areas. Although the *National Urban System Plan (2005–2020)* defines three megalopolises and 13 urban agglomerations that are comparable in terms of spatial coverage and differentiates the roles and levels of urban agglomerations, it overemphasizes the coastal development belt while overlooking the important role of inland development belts in promoting economic integration and coordinated regional development in China. This is contrary to the priorities defined in China's overall regional development strategy. The Chinese government should instead prioritize the further implementation of central and western regional development strategies such as the Western Development, Rise of Central Region and Northeast Revitalization strategies. Accordingly, it should build a strategic distribution of towns and cities according to the relative positions, sizes and strengths of urban agglomerations and categorize them into different levels—world-class and national/regional levels—so as to build multiple centers with different positions. In the meantime, the Chinese government should plan and deploy a system of major development belts from a medium- and long-term perspective while implementing a multi-center network development strategy.

With respect to long-term development, the Chinese government should conduct future land development and urbanization according to the aforementioned strategic, network-like distribution characterized by four latitudinal and four longitudinal belts (see Fig. 5.10 and Table 5.10). The four latitudinal belts refer to the following:

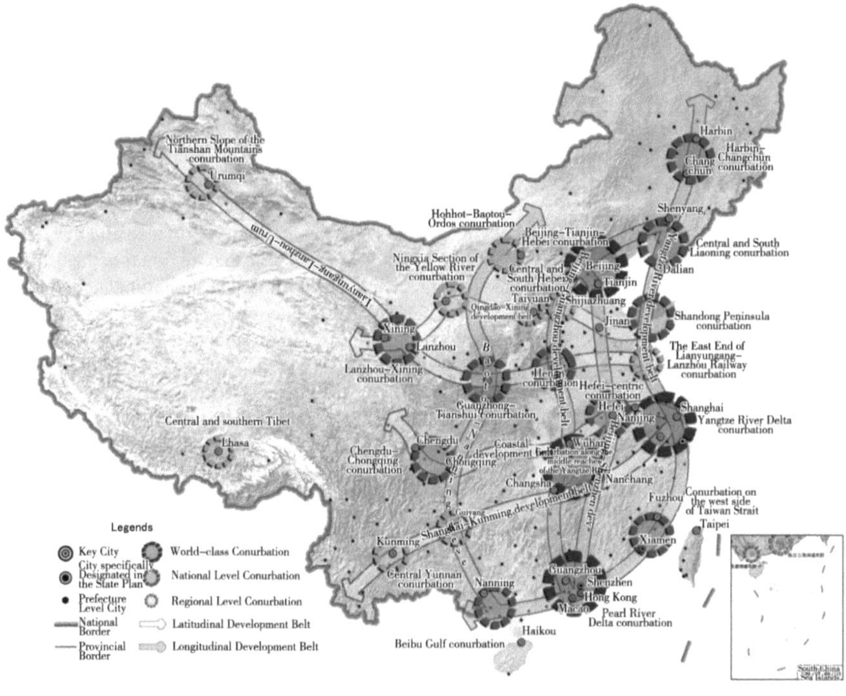

Fig. 5.10 China's spatial distribution strategy for urbanization by 2030 *Source* The author

Table 5.10 Land development and urbanization in China: four latitudinal and four longitudinal belts

Development belt	Belt name	Description
Four latitudinal belts	Yangtze River development belt	Spanning Shanghai, Nanjing, Hefei, Wuhan, Chongqing and Chengdu, this belt interconnects the Yangtze River Delta, Hefei, Middle Reaches of the Yangtze River and Chengdu-Chongqing urban agglomerations. There are 23 cities and over 100 development areas and high-tech industry zones along this belt, making it a nationally-important, people- and industry-agglomerated economic corridor and the most important link for enhancing economic ties between East and Central/West China
	Lianyungang-Lanzhou-Urumqi development belt	Spanning Lianyungang, Zhengzhou, Xi'an, Lanzhou and Urumqi, this belt interconnects the East End of the Lianyungang-Lanzhou Railway, Guanzhong-Tianshui, Lanzhou-Xining and Northern Slope of the Tianshan Mountains urban agglomerations. It runs through 10 Chinese provinces and autonomous regions from east to west and across 11 longitudinal railways, interconnecting the bordering region of Jiangsu, Shandong, Henan and Anhui Provinces, the rest of Henan Province and the Lanzhou-Urumqi economic belt. It is an economic development belt based on important railways
	Shanghai-Kunming development belt	Spanning Shanghai, Hangzhou, Nanchang, Changsha, Guiyang and Kunming, this belt interconnects the Yangtze River Delta, Middle Reaches of the Yangtze River, Central Guizhou and Central Yunnan urban agglomerations. It is an important link between east and southwest China as well as south and southeast Asian countries
	Qingdao-Xining development belt	Spanning Qingdao, Jinan, Shijiazhuang, Taiyuan, Yinchuan and Xining, this belt interconnects the Shandong Peninsula, Central and South Hebei, Taiyuan, Ningxia and Lanzhou-Xining urban agglomerations. By linking the BER, coal chemical industry bases in Shanxi Province and the northwestern Hohhot-Baotou-Yinchuan region together, this belt has become a nationally-important, latitudinal economic development belt. With rich energy and mineral resources, this belt serves as China's most important energy and raw materials production base. By enhancing the exploitation of energy and mineral resources, the coal, natural gas, oil and raw materials industries within this belt will offer greater support for national economic development, making it strategically significant for China's economic safety

(continued)

Table 5.10 (continued)

Development belt	Belt name	Description
Four longitudinal belts	Coastal development belt	Spanning Harbin, Changchun, Shenyang, Dalian, Qingdao, Shanghai, Hangzhou, Ningbo, Fuzhou, Xiamen, Shenzhen, Guangzhou and Nanning, this belt interconnects the Harbin-Changchun, Central and South Liaoning, Shandong Peninsula, East End of the Lianyungang-Lanzhou Railway, Yangtze River Delta, West Side of the Taiwan Strait, Pearl River Delta and Beibu Gulf urban agglomerations. Accounting for 14% of China's land area, over 40% of its population and nearly 70% of its economic aggregate, this is a development belt with the best developmental foundations, the greatest growth potential as well as the highest competitiveness and strategic significance in China
	Beijing-Guangzhou development belt	Spanning Shenyang, Beijing, Shijiazhuang, Zhengzhou, Wuhan, Changsha and Guangzhou, this belt interconnects the Central and South Liaoning, Beijing-Tianjin-Hebei, Central and South Hebei, Henan, Middle Reaches of the Yangtze River and Pearl River Delta urban agglomerations. Based on arteries such as the Beijing-Guangzhou Railway and Beijing-Zhuhai Highway, this belt covers nearly the entire Central China region while connecting the BER and Pearl River Delta regions. Consisting of 35 cities, it houses six important transport hubs that link all parts of China together and runs across the Haihe, Yellow, Yangtze and Pearl river systems
	Beijing-Shenzhen development belt	Spanning Beijing, Jinan, Hefei, Nanchang and Shenzhen, this belt interconnects the Beijing-Tianjin-Hebei, Hefei, Middle Reaches of the Yangtze River and Pearl River Delta urban agglomerations. With the BER and Beijing-Tianjin-Hebei urban agglomeration on the north end, the Pearl River Delta urban agglomeration on the south end and the Nanchang-Jiujiang Industrial Corridor as an emerging industry cluster in the middle, this belt abounds in mineral resources and is an important producer of grains, cotton and other agricultural products in China. It serves as an important bridge in helping China's developed coastal regions fuel the growth of central and western regions
	Baotou-Nanning development belt	Spanning Baotou, Xi'an, Chongqing, Guiyang and Nanning, this belt interconnects the Hohhot-Baotou-Ordos, Ningxia Section of the Yellow River, Guanzhong-Tianshui, Central Guizhou, Chengdu-Chongqing and Beibu Gulf urban agglomerations. With the Hohhot-Baotou-Ordos region on the north end, the Beibu Gulf region on the south end and Guanzhong, Chengdu and Chongqing in the middle, this belt intersects with the latitudinal Lianyungang-Lanzhou-Urumqi and Yangtze River development belts

Source The author

1. Yangtze River development belt (Shanghai, Nanjing, Hefei, Wuhan, Chongqing, Chengdu);
2. Lianyungang-Lanzhou-Urumqi development belt (Lianyungang, Zhengzhou, Xi'an, Lanzhou, Urumqi);
3. Shanghai-Kunming development belt (Shanghai, Hangzhou, Nanchang, Changsha, Guiyang, Kunming); and
4. Qingdao-Xining development belt (Qingdao, Jinan, Shijiazhuang, Taiyuan, Yinchuan, Lanzhou, Xining).

The four longitudinal belts refer to the following:

1. Coastal development belt (Harbin, Changchun, Shenyang, Dalian, Qingdao, Shanghai, Hangzhou, Ningbo, Fuzhou, Xiamen, Shenzhen, Guangzhou, Nanning);
2. Beijing-Guangzhou development belt (Shenyang, Beijing, Shijiazhuang, Zhengzhou, Wuhan, Changsha, Guangzhou);
3. Beijing-Shenzhen development belt (Beijing, Jinan, Hefei, Nanchang, Shenzhen); and
4. Baotou-Nanning development belt (Baotou, Xi'an, Chongqing, Guiyang, Nanning).

Building such a strategic distribution and implementing a multi-center network development strategy will facilitate economic integration and coordinated regional development in China, thus establishing a pattern of land development and urbanization with high levels of safety, orderliness, balance and efficiency.

References

Cai, Y. F., & Zhang, C. W. (2012). A new view of interregional gaps: research on the mismatch between population and industry distributions. *China Industrial Economics, 5,* 31–43.

Fujita, M., Krugman, P. R., & Venables A. J. (2001). *The spatial economy: Cities, regions, and international trade*. MIT press.

Guan, X. L., Fang, C. L., & Luo, K. (2012). Space field energy-based assessment on interregional gaps in economic development across China. *Scientia Geographica Sinica, 32*(9), 1055–1065.

Hu, H. Y. (1935). Population distribution in China. *Acta Geographica Sinica, 2*(2), 33–74.

Krugman, P. (1991). Increasing returns and economic geography. *Journal of Political Economy, 99*(3), 483–499.

Liu, Y. (2014). On the efficiency of land utilization and potential of land conservation in China. In Wei, H. K (Ed.), A new road to urbanization with Chinese characteristics. Beijing: Social Sciences Academic Press (China).

Lu, M., & Chen, Z. (2004). On urbanization, city-biased economic policies and urban-rural income gap. *Economic Research Journal, 6,* 50–58.

National Bureau of Statistics. (2014, May 12). *Migrant workers monitoring survey report 2013.*

Wei, H. K. (2003). On how industry clusters relate to competitiveness. *Economic Management Journal, 6,* 4–11.

Wei, H. K. (2009a, June 30). Beyond disorderly development: Urban development must be within control. *China Economic Herald*.

Wei, H. K. (2009b). A new land development strategy of China in a new stage. *Green Leaf, 9,* 38–43.
Wei, H. K. (2014). *A new road to urbanization with Chinese characteristics*. Beijing: Social Sciences Academic Press (China).
Wei, H., Wang, Y., & Bai, M. (2014). *The micro-analysis of regional economy in China: A perspective of firm relocation*. Singapore: World Scientific.
Wei, H. K., et al. (2011). *On China's regional policies—assessment and outlook*. Beijing: Economic Management Press.
Wen, K., Zhu, E. J., et al. (2013). *Annual report on Beijing-Tianjin-Hebei metropolitan region development (2013)*. Beijing: Social Sciences Academic Press (China).
Wu, B., Che, X. Z., Chen, X. D., & Xie, L. S. (2012). On Shenzhen's aquatic environment: Capacity and carrying capacity. *Research of Environmental Sciences, 25*(8), 953–958.
Yang, W. M. (2008). Establish main functional areas and improve the pattern of land development. *Economic Review, 5,* 17–21.
Zhang, Y. L. (2011, July 26). Such a limit must never be exceeded: The land development intensity in the Pearl River Delta is much higher than the globally accepted 30% limit. *China Land and Resources News*.

Chapter 6
Administrative Levels and Size Growth of Chinese Cities

Urbanization is mainly manifested in the growth of city sizes. Cities in China are distinctive from those in other countries in that they have different administrative level classifications. A Chinese city's size and growth has much to do with its administrative level. A higher administrative level can result in exponential increases in urban population and land area used for urban development. Cities at higher administrative levels tend to grow faster than those at lower administrative levels in terms of population growth and land area development. Such differences are significantly affected by Chinese government allocation of resources in favor of administrative centers. An important reason for such inter-city differences lies in differentiated power settings, resources allocation and institutional configurations for cities at different administrative levels. In order to facilitate the well-coordinated development of Chinese towns and cities of varying sizes, the Chinese government must radically change the current administrative center-biased way of allocating resources and reduce the political weight of administrative levels. The Chinese government should instead base city size control upon the area's overall carrying capacity and fuel the growth of economically-developed towns by granting them more administrative powers, thus increasing inter-city equality with respect to development opportunities, rights and benefits.

6.1 Why Has China's City Size Control Policy Failed?

The Chinese government has long focused its urban development guidelines and policies on controlling the size of larger cities. Back in the late 1950s and 1960s, former Chinese leader Mao Zedong emphasized that the Chinese government would control the size of larger cities and develop smaller cities, in line with Mao's estimation of the international situation (Zhou 1990). The idea of controlling the size of larger cities produced a profound influence on China's subsequent guidelines and policies regarding urban development. From the 3rd National Urban Work Conference in March

H. Wei, *Urbanization in China*, Research Series on the Chinese Dream and China's Development Path, https://doi.org/10.1007/978-981-13-1408-7_6

1978 to the *Summary of the National Urban Planning Work Conference* released by the State Council in December 1980, the Chinese government consistently made clear its commitment to controlling the size of larger cities while developing smaller ones. This policy was strengthened by the *Urban Planning Law of the People's Republic of China,* which became effective in April 1990 but is now no longer valid. After 2000, the Chinese government announced a new guideline of promoting coordinated development of towns and cities of varying sizes. Nonetheless, it has yet to radically change the policy of controlling the size of larger cities. The outline of the 10th Five-Year Plan outlined measures that should be taken to prevent city sizes from increasing in an unreasonable manner; the outline of the 12th Five-Year Plan stated that the excessive area expansion of very large cities must be prevented. In addition, it was made clear at both the 3rd Plenary Session of the 18th CPC Central Committee and the Central Urbanization Work Conference that the population sizes of very large cities must be kept under tight control. The *National New Urbanization Plan (2014–2020)* released in March 2014 further stated that the population sizes of larger cities with an urban-area population over 5,000,000 people must be kept under tight control. As these examples demonstrate, although China's guidelines regarding urban development have changed since its reform and opening, policies regarding control over the size of larger cities remain unchanged except for an adjustment to wording: "... control the size of large cities" has been replaced with "... maintain tight control over the population size of very large cities".

Controlling the size of larger cities mainly refers to control of population size and land area used for urban development, not control of economic growth. According to policies implemented since reform and opening, China controls the size of larger cities mainly by three means. First is control over urban planning. China approves general urban plans on a per-level basis. Under the *Urban Planning Law of the People's Republic of China*, municipalities directly under the Central Government, provincial or regional capitals, other cities with an urban population of over 1,000,000 people and cities identified by the State Council must submit their general plans to the Central Government for approval. In the *Notice on Enhancing Urban Planning* released in May 1996, the State Council announced that larger cities with a non-agricultural population of over 500,000 people must submit their general plans for approval and that they must first report their urban population size and land area used for urban development to the competent national authorities for verification. Effective in January 2008, the *Urban and Rural Planning Law of the People's Republic of China* provides that municipalities directly under the Central Government, provincial or regional capitals and cities identified by the State Council must submit their general plans to it for approval. Up to now, general urban plans approved by the State Council all include targets for controlling urban population sizes and land area used for urban development.

The second means of control is control over urban *hukou*. China has long controlled the population size of larger cities through tight resident registration policy. Although towns and small/medium-sized cities have loosened control over *hukou* over the past few years as part of the overall *hukou* system reform effort, large cities, especially very large cities, still adhere to a tight resident registration policy.

The ongoing national level *hukou* system reform effort is being carried out by first addressing less difficult issues and starting from smaller cities. It remains necessary for large cities to determine reasonable requirements for resident registration and for very large cities to keep tight control over population size. Today in Beijing, Shanghai and other very large cities, control over *hukou* remains the primary means of controlling urban population growth.

The third means of control is control via industry shift. The outline of the 11th Five-Year Plan called on very large cities to adjust their industrial structures and control population growth by economic means. With rapid suburbanization in larger Chinese cities since the 1990s, these cities have been racing to prioritize the growth of the tertiary sector over the secondary sector and commercial development projects over residential ones. In addition, these cities have been guiding industries in relocating from central districts to the outskirts by such means as relocation under administrative instructions, land replacement and the setting of entry/exit standards (Wei et al. 2014). Such industry shift is largely part of suburbanization and, hence, is carried out within a limited area. This practice has caused the further size expansion of larger cities.

China's policies aimed at controlling the size of larger cities have failed to work as expected. Such failure is demonstrated by the polarization trend which has appeared amid China's urbanization in recent years (Wei 2014). Some scholars even refer to the current population growth of very large Chinese cities as "explosive growth" and believe that its consequences "may also be explosive or even destructive" (Ding 2011). It is safe to say that the rapidly growing sizes of larger Chinese cities are against the desires and regulatory efforts of the Central Government. Why has its control policy failed? Why are larger Chinese cities still expanding rapidly despite these control policies? Scholars have explored both questions from differing angles. Some scholars believe that larger cities taking the lead in number and size growth is a natural law (Hu et al. 1985; Hu 1986; Gao 1991), and that efforts to control the size of larger cities contravenes this law and is both one-sided (Wang 1985) and paradoxical (Wang 2014). Proponents of this view object to controlling the size of larger cities and advocate prioritizing the development of larger cities. Other scholars explain the unreasonably rapid growth of larger cities from the perspective of economies of agglomeration, believing that the rapid growth is a result of the concentration of people, resources and other factors of production into larger cities due to their economies of scale and of agglomeration and functions that are too large in number (Rao 1989; He 1991). Such concentration leads to excessive agglomeration, which in turn will cause severe diseconomies of agglomeration once the concentration exceeds the overall carrying capacity of a city, at which point "urban diseases" will prevail.

Economies of agglomeration are an important, but not the primary, driving force for the rapid growth of larger cities. That the sizes of these cities have grown rapidly in the earliest stage of urbanization cannot be explained fully by economies of agglomeration. A third view focuses on how administrative levels have affected the growth of city sizes in different dynasties. Early studies generally regard administrative levels of cities as an important determinant of their size (Ma 1998; Zhang 2000; Zhou 2001). Nonetheless, Cheng (2007) questioned this view after he studied Chinese cities in the Qing Dynasty, believing that there is a weak correlation between administrative level and city size, namely, the latter is not decided by the former. Since China does not yet have a fully developed market economy, the market plays a relatively limited role in socioeconomic development, which is significantly affected by political factors and administrative forces. Consequently, administrative levels of cities will necessarily produce a significant influence on the growth of city size. Unfortunately, however, existing studies have all focused on the periods before the People's Republic of China was founded and, accordingly, fall under urban history. Few scholars have conducted in-depth research on how the administrative levels of Chinese cities relate to their size growth today.

Outside China, the transformation of towns into cities is typically governed by citizens themselves. Populous, physically built-up cities in other countries differ from each other only in size, not in administrative level (Wang 2013). These cities fundamentally have equal access to development opportunities and rights regardless of their size. In comparison, Chinese cities are generally formed on a system-wide basis. Existing Chinese cities are all administrative areas that include urban and rural areas. Moreover, Chinese cities are ranked at different levels in the administrative hierarchy, which results in wide inter-city differences in political status as well as legislative and administrative powers. Towns and cities at lower levels must follow the leadership of cities at higher levels. Such an administrative hierarchy built upon administrative forces is a primary characteristic that distinguishes Chinese cities from foreign ones. Because of the presence of this administrative hierarchy, Chinese cities differ widely from each other in terms of political status and access to resources, rights/benefits and even development opportunities. The combination of inter-city inequality caused by differences in administrative level, biased resource allocation in favor of administrative centers, and market forces together produce complex, mutually enhancing effects that result in the excessive expansion of cities at higher administrative levels and failure of the Chinese government's size control policy. Among these factors, it is the bias toward administrative centers and the administrative hierarchy that is the primary cause and driving force for the rapid size growth of larger Chinese cities, despite the government's size control policy.

6.2 Correlation Between Administrative Levels and Size Growth of Chinese Cities

China's administrative hierarchy in terms of urban management is epitomized by the administrative levels assigned to Chinese cities. Chinese towns are de facto cities, according to internationally accepted standards, and are thus are also analyzed in this study. According to the administrative hierarchy, Chinese cities can be categorized into municipalities directly under the Central Government, cities at the prefecture level, cities at the county level and towns. By the end of 2012, there were four municipalities directly under the Central Government, 285 cities at the prefecture level, 368 cities at the county level and 19,881 towns across China excluding Hong

Kong, Macao and Taiwan. Given the importance of administrative levels and government seats, Chinese cities can be further categorized into municipalities directly under the Central Government, cities at the sub-provincial level, general provincial capitals, general cities at the prefecture level, cities at the county level, county seats and general towns.

In the administrative hierarchy of cities, Beijing, Tianjin, Shanghai and Chongqing—the four municipalities directly under the Central Government—are Tier 1 cities. With an administrative level equal to that of 22 provinces and five autonomous regions, these four cities hold legislative and administrative powers at the provincial level, with certain members of the Political Bureau of the CPC Central Committee doubling as the heads of their respective CPC municipal committees. Cities at the prefecture level can be further categorized into cities at the sub-provincial level, general provincial capitals and general cities at the prefecture level depending on their administrative levels and whether they are provincial or regional capitals. Of these, the State Commission Office for Public Sector Reform (SCOPSR) has identified fifteen cities at the sub-provincial level, including Shenzhen, Guangzhou, Xiamen, Hangzhou, Ningbo, Chengdu, Nanjing, Changchun, Shenyang, Dalian, Harbin, Jinan, Qingdao, Wuhan and Xi'an. These are considered Tier 2 cities within the administrative hierarchy. Although they also are administered by their respective provinces, these cities hold administrative powers at the sub-provincial level in general and even have some powers at the provincial level with regard to socioeconomic development planning. The other 17 general provincial capitals are considered Tier 3 cities within the administrative hierarchy. Despite their lack of higher administrative powers, these cities see certain members of the CPC provincial standing committees double as the heads of their respective CPC municipal committees. Moreover, these cities have advantages as provincial or regional capitals with power to make local laws.

There are 253 general cities at the prefecture levels, including larger municipalities,[1] which are considered Tier 4 cities within the administrative hierarchy. There are also certain differences among cities at the county level in China. For example, some provinces have cities at the sub-prefecture level or cities at the county level administered directly by the provincial government. These cities are granted additional powers regarding socioeconomic management. Since this practice differs from province to province and accurate, detailed statistics are unavailable, cities at the county level are not further categorized herein and are all viewed as Tier 5 cities within the administrative hierarchy. Of all towns in China, county seats tend to have enormous advantages—certain members of the CPC county standing committees usually double as the heads of their respective CPC town committees and they have advantages in attracting investment and accessing public resources. Therefore, county seats are ranked as Tier 6 within the administrative hierarchy, and all other Chinese towns are at Tier 7. In recent years, a number of Chinese towns have

[1] Larger municipalities refer to the seats of 27 provincial or autonomous region governments, special economic zones (i.e., Shenzhen, Zhuhai, Xiamen and Kashgar) and 18 other larger municipalities approved by the State Council. These cities hold the power of making local laws and regulations.

been granted additional administrative powers, and are referred to as pilot towns with additional powers, towns administered directly by the municipal government or towns at the sub-county level. Nonetheless, since relevant local standards differ from one another and accurate statistics are unavailable, Chinese towns at Tier 7 are not further categorized herein.

City size can be measured by urban population and land area used. With data released by the MOHURD, this study uses town/city area or county seat population sizes to measure urban population and the size of built-up areas and land used for urban development to measure land area used by cities. The administrative hierarchy and average size of Chinese cities in 2012 are shown in Table 6.1. As can be seen from this table, the numbers of Chinese cities at different administrative levels are roughly in a pyramid-shaped distribution, while their average size is in an upside-down pyramid-shaped distribution. The higher (or lower) the administrative level, the smaller (or larger) the number of cities. With respect to average city size, the higher (or lower) the administrative level of a city, the larger (or smaller) the urban population size and land area used. Take urban population for example. In 2012, 1624 county seats in China had an average population of 92,000 people. Cities at the county level, general cities at the prefecture level, general provincial capitals, cities at the sub-provincial level and municipalities directly under the Central Government were 2.3, 7.4, 27.3, 51.8 and 161.2 times as high as county seats in this regard respectively. This also held true for the amount of land area used by cities at different administrative levels, only with slightly narrower inter-city gaps. This is because cities at higher administrative levels have higher population density and efficiency of land utilization thanks to larger city sizes. In the administrative hierarchy of Chinese cities, towns have an average population of 10,000 people and an average amount of 2.2 km^2 of developed land. They are far behind county seats in both dimensions and are unable to meet the required minimum efficient size of 30,000 people (Wei 2010). General towns are even smaller in average size, with an average of 1.2 km^2 of developed land, or only about 10% of that of county seats.

From an average size perspective, the population size and amount of land use of Chinese cities are highly correlated with their administrative levels. This study uses data from 2010–2012 to plot how the average size of Chinese cities relates to their administrative levels, as shown in Fig. 6.1. The data shows that both urban population and land area used increase exponentially as the administrative levels of Chinese cities rise from Level 7 to Level 1. The three fitted exponential functions each have a coefficient of determination (R^2) of over 0.96, which suggests that the corresponding curve is well fitted, regardless of urban-area population, built-up area size, and land used for urban development. To further study the correlation between the size and administrative level of a city, data for approximately 656 cities from the *China Urban Construction Statistical Yearbook* (2012) was analyzed. County seats and general towns were not included due to the unavailability of specific data. Since the administrative level of a city falls under ordered categorical variables, this study used Spearman's rank correlation coefficient for analysis. The result showed that the correlation coefficient was −0.744 between the administrative level and urban-area population of a city, −0.752 between the administrative level and built-up area

Table 6.1 Administrative hierarchy and average sizes of Chinese cities, 2012

Level 级	Level name	Count	Urban-area population[c]		Built-up area		Land used for urban development[d]	
			Average size (in million people)	Times (1 for county seats)	Average size (km^2)	Times (1 for county seats)	Average size (km^2)	Times (1 for county seats)
1	Municipality directly under the central government	4	14.830	161.2	1008.4	87.7	1482.7	138.6
2	City at the sub-provincial level	15	4.769	51.8	489.9	42.6	476.2	44.5
3	General provincial capital	17	2.514	27.3	235.7	20.5	236.5	22.1
4	General city at the prefecture level	252[a]	0.682	7.4	82.5	7.2	82.0	7.7
5	City at the county level	367	0.210	2.3	25.5	2.2	24.4	2.3
6	County seat	1624	0.092	1.0	11.5	1.0	10.7	1.0
7	General town[b]	15,563	–	–	1.2	0.1	–	–

Notes [a]Sansha City is excluded; [b]The total number is the number of all towns counted, or 17,187, minus the number of county seats; [c]Temporary population is included and, for county seats, urban-area population refers to population in the county seat; [d]Data for the cities of Shenzhen and Jieyang, Guangdong Province and Mangshi City, Yunnan Province is unavailable and therefore is replaced with the size of their developed land areas; data for Jiyuan City, Henan Province is erroneous and has been corrected

Source: *China Urban Construction Statistical Yearbook* (2012) and *China Urban-Rural Construction Statistical Yearbook* (2012)

size of a city, and −0.734 between the administrative level and land area used for urban development, which were all significant at level 0.01 with a two-tailed test. If we used the Kendall tau-b coefficient for analysis, the correlation coefficient values would drop slightly but would be all significant at level 0.01. Of these, the correlation coefficient was −0.620 between the administrative level and urban-area population of a city, −0.627 between the administrative level and built-up area size of a city, and −0.610 between the administrative level and land area used for urban development. This indicates that the size and administrative levels of Chinese cities are significantly correlated.

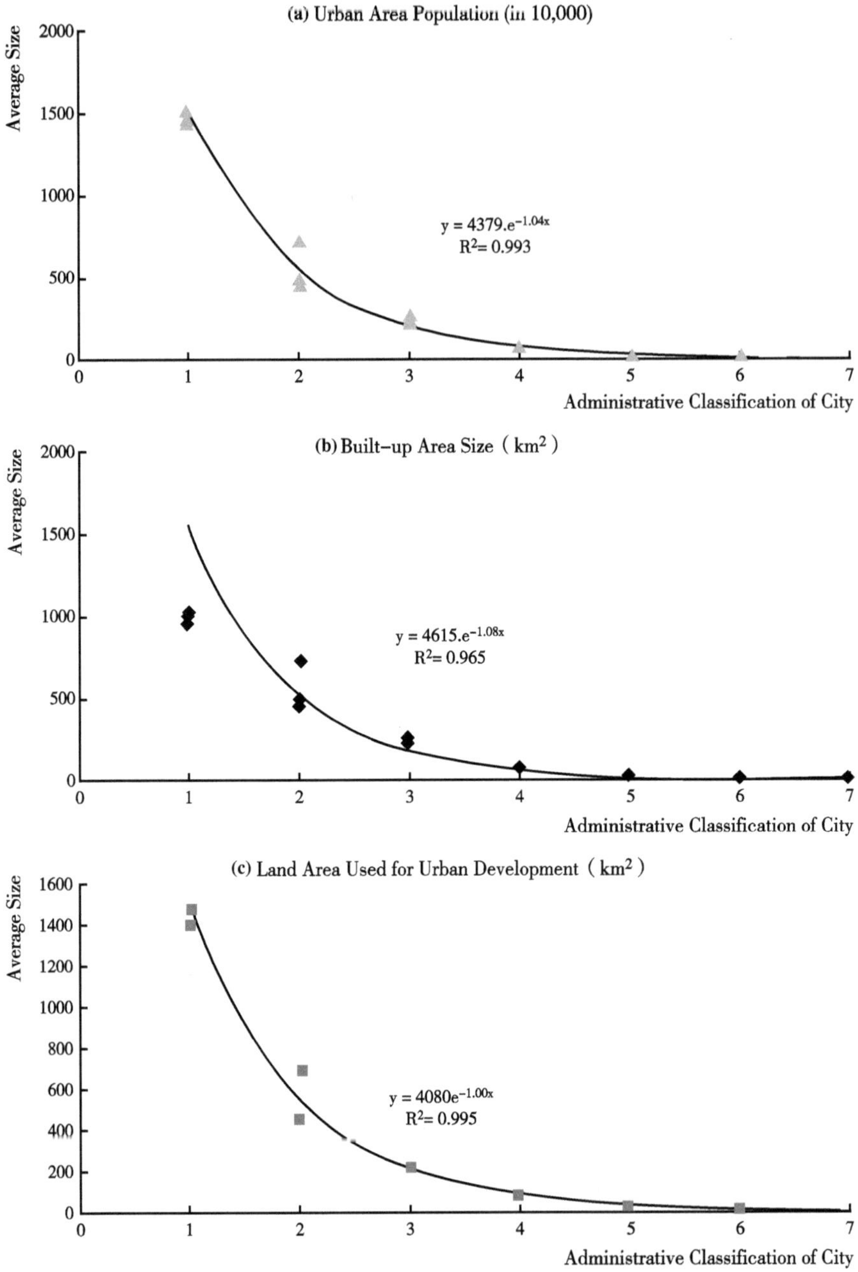

Fig. 6.1 Scale increase of chinese cities by administrative classification Source: *China Urban Construction Statistical Yearbook* (2010–2012) and *China Urban-Rural Construction Statistical Yearbook* (2012)

Cities at higher administrative levels are larger in size mainly because of the interplay of original size and recent rapid expansion. Historically, municipalities directly under the Central Government, cities at the sub-provincial level and provincial capitals are the most ancient cities in China, with their existing size resulting from continuous expansion and accumulation. Most of these cities were already regional administrative centers; others were identified as such centers or raised to higher administrative levels due to their larger sizes, economic significance and the requirements of China's reform and opening. It can be seen from Table 6.2 that the size and administrative level of Chinese cities are clearly correlated and remained relatively stable between 2002 and 2012. Ultra-large cities with populations over 4,000,000 people are mostly municipalities directly under the Central Government or cities at the sub-provincial level. Very large cities with populations of 1,000,000–4,000,000 people are mostly general cities at the prefecture level, along with some cities at the sub-provincial level and provincial capitals. Large cities each with populations between 500,000–1,000,000 people are mostly general cities at the prefecture level. Medium-sized cities with populations between 200,000–500,000 people are all general cities at the prefecture level or cities at the county level with the exception of Lhasa. Almost all small cities with populations fewer than 200,000 people are cities at the county level. According to relevant analysis, the Kendall tau-b coefficient and Spearman's rank correlation coefficient between the size and administrative level of 656 Chinese cities were 0.666 and 0.711 respectively in 2012, both of which were significant at level 0.01. This indicates that the size and administrative level of Chinese cities are significantly correlated.

To study increases in the average size of Chinese cities at different administrative levels, we selected the years 2006 and 2012 for comparative analysis. These years were selected mainly because the number of Chinese cities remained unchanged in this period, the administrative hierarchy remained fundamentally stable, and consistent statistical metrics were available. This allows for better comparability. According to our analysis, urban-area populations in municipalities directly under the Central Government, cities at the sub-provincial level and general provincial capitals grew at a much faster rate than general cities at the prefecture level and cities at the county level between 2006 and 2012. Of these, municipalities directly under the Central Government grew 5.1 and 3.3 times faster than general cities at the prefecture level and cities at the county level respectively; general provincial capitals grew 5.5 and 3.6 times faster than general cities at the prefecture level and cities at the county level respectively; county seats also grew much faster than general towns in terms of population growth (see Table 6.3). This indicates that during this period, China's urban population was concentrating mainly into municipalities directly under the Central Government, cities at the sub-provincial level and provincial capitals, all of which are at higher administrative levels, as well as county seats, which are administrative centers of counties. General cities at the prefecture level, cities at the county level and general towns witnessed slower population growth. In contrast, land size used for urban development increased under a different pattern. All Tier 2–4 cities at the prefecture level expanded rapidly, whereas Tier 1 municipalities directly under the Central Government and Tier 5 cities at the county level expanded relatively slowly.

Table 6.2 Correlation between size and administrative level of Chinese cities

Year	Administrative/size	Total	>4.0 million	1.0–4.0 million	0.5–1.0 million	0.2–0.5 million	<0.2 million
2002	Municipalities directly under the central government	4	*4*	–	–	–	–
	Cities at the sub-provincial level	15	*5*	*10*	–	–	–
	General provincial capitals	17	–	*13*	3	1	–
	General cities at the prefecture level	239	–	*56*	*83*	*87*	13
	Cities at the county level	384	–	4	17	*113*	*250*
2006	Municipalities directly under the Central Government	4	*4*	–	–	–	–
	Cities at the sub-provincial level	15	*6*	*9*	–	–	–
	General provincial capitals	17	–	*13*	3	1	–
	General cities at the prefecture level	251	1	*34*	*83*	*103*	30
	Cities at the county level	369	–	1	7	*120*	*241*
2012	Municipalities directly under the Central Government	4	*4*	–	–	–	–
	Cities at the sub-provincial level	15	*7*	*8*	–	–	–
	General provincial capitals	17	2	*14*	–	1	–
	General cities at the prefecture level	253	1	*40*	*96*	*101*	15
	Cities at the county level	367	–	–	4	*169*	*194*

Note Urban population figures are used for 2002 and urban-area population (including temporary population) figures are used for 2006 and 2012

Source: *China Urban Construction Statistics Yearbook* (2002) and *China Urban Construction Statistical Yearbook* (2006 and 2012)

County seats expanded much faster than general towns, with built-up areas in the former expanding 7.9 times faster than those of the latter. Since all municipalities directly under the Central Government are already very large, they are now facing constraints in terms of overall carrying capacity, especially the resource and environmental carrying capacity. Despite these constraints, these cities still expanded faster than cities at the county level in terms of land size used for urban development, and much faster than general towns in terms of built-up areas during this period. In addition, from the perspective of growth elasticity for built-up areas and land used for urban development, general cities at the prefecture level saw values of 6.58 and 5.85 respectively in this period, which were far beyond the reasonable range of this coefficient.[2] Cities at the sub-provincial level also saw the growth elasticity of built-up area and land used for urban development reach 2.65 and 3.37 respectively, suggesting that land size used for urban development was increasing rapidly in these cities. This indicates that most Chinese cities at the prefecture level have the impetus and impulse to expand in size amidst ongoing and intense inter-city competition.

6.3 Consequences of Administrative Center Bias for City Size Growth

City size growth reflects the rise and fall of a city. In ancient times, Chinese cities largely served political functions, and their size growth was mainly affected by political, military and other factors. Since modern times, the political functions of a city remain critical to its development. Some cities have gradually waned due to the weakening of such functions. In the highly centralized planned economy after the People's Republic of China was founded, administrative centers at various levels naturally developed and grew in size at rapid rates due to the government's dominance in resources allocation. Despite the current transition to a market economy, the established administrative hierarchy remains deeply rooted and continues to exert significant influence on urban development. The size expansion of cities at higher administrative levels is facilitated by a power and institutional framework that favors administrative centers. The influence of administrative forces will continue to intensify the urban concentration of people, industries and factors of production in cities with higher administrative levels due to economies of agglomeration, in turn resulting in rapid expansion of city sizes.

The ongoing rapid size expansion of Chinese cities at higher administrative levels is a result of the centralized functions of administrative centers at higher administrative levels, the advantageous position these centers hold with regard to government resources allocation, and market forces. In this study, bias toward administrative centers refers to government allocation of powers and resources and other institutional configurations in favor of administrative centers, especially cities at higher administrative levels. Such a bias is reflected in the granting of power, resources, and

[2]It is generally believed that this coefficient should be around 1.12 (Yang 1998).

Table 6.3 Growth of average city size in China, 2006–2012

Level	Level name	Count		Average Size Growth, 2006–2012 (%)			Growth elasticity coefficient[d]	
		2006	2012	Urban-area population	Built-up area size	Land used for development[c]	Built-up area size	Land used for development
1	Municipalities directly under the central government	4	4	30.4	22.8	22.4	0.75	0.74
2	Cities at the sub-provincial level	15	15	13.9	36.8	46.9	2.65	3.37
3	General provincial capitals	17	17	32.8	44.5	42.6	1.36	1.30
4	General cities at the prefecture level	251	252[a]	6.0	39.5	35.1	6.58	5.85
5	Cities at the county level	369	367	9.1	28.1	21.9	3.10	2.41
6	County seats	1,635	1,624	26.3	42.6	30.5	1.62	1.16
7	General towns	16,017	15,563	10.4[b]	5.4	–	–	–

Notes [a]Sansha City is excluded; [b]The average population (incl. temporary population) in built-up areas of all towns is used; [c]Data about land used for development in the cities of Shenzhen and Shantou, Guangdong Province in 2006 is unavailable and therefore is replaced with the sizes of their built-up areas; data about the cities of Yangjiang and Jieyang, Guangdong Province is unavailable and relevant means for 2005 and 2007 are used for estimation; data about Shanghai is unavailable and the percentages of its built-up area and land used for development in 2007 are used for estimation; data for 2012 is processed in the same way as in Table 6.1; [d]It refers to the ratio between the growth rates of land used for urban development and of urban-area population

Source: *China Urban Construction Statistical Yearbook* (2006 and 2012) and *China Urban-Rural Construction Statistical Yearbook* (2012)

policy support to centers and cities at higher administrative levels, while overlooking and discriminating against other Chinese cities. Biased allocation of resources is an extension of the administrative hierarchy into urban development and management. It contravenes the principles of fairness and impartiality in the market economy, has led to unfair inter-city competition and worsened inequality in urban development. Biased resource allocation has become a major obstacle to coordinated urban development, and is a major reason that the growth rate of city size does not completely depend on a city's competitiveness. Cities at higher administrative levels keep expanding thanks to their greater administrative powers and availability of resources, whereas towns and cities at lower administrative levels are disadvantaged and even declining due to deprivation of certain resources and rights. As a result of this bias, administrative centers, China's capital, municipalities directly under the Central Government and provincial or regional capitals have long enjoyed centralized functions, allowing them to become regional or even national centers in such fields as politics, transport, culture/sports, sciences/technologies/education, commerce/logistics, finance, industrial manufacturing and tourism/recreation. This has caused a steep increase in city sizes and an overburden on some cities. Today, China's provincial or regional capitals tend to be the primary cities and economic centers in relevant provinces and autonomous regions. With such a bias toward administrative centers, the relocation of provincial governments has generally led to economic changes in relevant cities, more specifically the relative decline of the former provincial capital compared to the rise of the new capital (Wei 2014). Table 6.4 shows the wide gap in urban population between former and current provincial capitals caused by the relocation of provincial governments after the founding of the People's Republic of China. The city of Kaifeng, Henan Province, in particular, has long lost its prosperity since the provincial government was moved to the city of Zhengzhou. In contrast to the slow urban population growth in Kaifeng, Zhengzhou has since expanded sharply. By 2012, Zhengzhou was already 6.55 times larger than Kaifeng in terms of urban-area population.

The bias toward administrative centers impacts the growth of city size mainly by three means:

Power settings: Under a centralized system of government, the Central Government holds unlimited powers, and administrative centers and nearby cities maintain prosperity by receiving more resources and political support (Liu et al. 2005; Zhao 2006; Deng 2011). Thanks to market oriented reforms since China began reform and opening up, the market has become more important in allocating resources while the role of the government has significantly declined. Nonetheless, the government still plays a critical role in economic development. There are wide differences between cities at higher administrative levels (municipalities directly under the Central Government, cities at the sub-provincial level and larger municipalities) and those at lower levels (general cities at the prefecture level, cities at the county level and towns) with respect to legislative, administrative and socioeconomic management powers. For cities at higher administrative levels, more administrative and managerial powers mean more development opportunities and stronger ability to mobilize and allocate resources. For towns and cities at lower administrative levels, limited powers mean the loss of certain development opportunities and inability to mobilize particular resources. In particular, during China's financial system reform over the past few years, there has been a severe imbalance between responsibilities and rights/benefits in Chinese cities at different administrative levels. Local governments have been given more responsibility over public affairs but the Central Government has assumed more financial powers than ever before. Cities at the county level and towns at the bottom of the administrative hierarchy are burdened with responsibility over public affairs but lack sufficient financial and managerial powers. Granting pow-

Table 6.4 Relocation of provincial/regional governments versus city size growth after the founding of the People's Republic of China

Province	Relocation			Urban population, 2000 (in million people)			Urban-area population, 2012 (in million people)		
	Year	From	To	Former capital (A)	Current capital (B)	B/A	Former capital (A)	Current capital (B)	B/A
Hebei	1968	Baoding	Shijiazhuang	0.8662	1.6100	1.86	1.2113	2.5101	2.07
Inner Mongolia	1953	Zhangjiakou	Hohhot	0.7142	1.0600	1.48	0.8680	1.8901	2.18
Jilin	1954	Jilin	Changchun	1.7934	2.9283	1.63	1.2793	3.6100	2.82
Heilongjiang	1954	Qiqihar	Harbin	1.4365	3.3820	2.35	1.0885	4.3061	3.96
Henan	1952	Kaifeng	Zhengzhou	0.7725	2.7247	3.53	0.8886	5.8166	6.55
Guangxi	1950	Guilin	Nanning	0.6430	1.3564	2.11	0.8220	2.4808	3.02

Note Urban or urban-area population includes temporary population
Source: *China Urban Construction Statistics Yearbook* (2000), *China Urban Construction Statistical Yearbook* (2012), and so forth

ers to municipal governments depending on their administrative level is conducive to the development and size expansion of larger cities at higher administrative levels but detrimental to those of towns and cities at lower levels, especially ones in the countryside.

Resource allocation: Administrative centers are the seats of local governments. In addition to government organizations, government-controlled resources such as higher education organizations, research institutions, financial institutes, state-owned enterprises, large hospitals and cultural/art organizations are highly concentrated in administrative centers at higher levels. Foreign-invested businesses, private companies, quality human resources and individuals who can influence decision makers are also concentrated in these administrative centers for their proximity to centers of power and access to high-quality public services. Since administrative centers have power and decision-making authority, cities at higher administrative levels are generally able to receive more resources and support from upper-level governments. The geographic proximity effect and concentration of political resources and social capital serve as the main reason why the Central Government is inclined to invest more resources in the national capital, municipalities directly under the Central Government and cities at the sub-provincial level, why provincial/regional governments invest more resources in provincial/regional capitals and why cities at the prefecture level invest more resources in central urban areas. The higher the administrative level of a city, the higher the average investment in municipal facilities per capita, for example. In 2006, municipalities directly under the Central Government, cities at the sub-provincial level, general provincial capitals and general cities at the prefecture level were 4.22, 2.72, 1.98 and 1.30 times higher than cities at the county level respectively in terms of the average investment in municipal facilities per capita. Such inter-city gaps have narrowed over the past few years, but municipalities directly under the Central Government, cities at the sub-provincial level and provincial capitals remain over twice as high as cities at the county level in this regard (see Table 6.5). Chinese towns fare even lower than cities in this regard. Consequently, there are wide gaps in infrastructure and provision of public services among Chinese cities at different administrative levels. According to 2012 data, cities at higher administrative levels scored higher in terms of municipal facilities than cities at lower levels, not to mention county seats and general towns (see Table 6.6). Since 1999, in particular, state-owned capital controlled by the government has been flowing into cities at higher administrative levels and with more political resources (Deng 2011). This indicates that, under the current pattern of resource allocation, a city with a higher administrative level and closer proximity to centers of power are able to receive more opportunities and resources and are thus better able to grow in size.

Institutional configurations: Institutional configurations are a key factor for urban growth. Given its lack of experience, China has adopted a trial-and-error approach in carrying out numerous important reforms and market opening initiatives. China has followed a model of first implementing certain programs in a small number of regions that meet particular requirements, before spreading the acquired experience to the rest of China. In the process of China's reform and opening, cities at higher administrative levels have generally been better able to secure favorable

Table 6.5 Inter-city Comparison of Average Investment in Municipal Facilities per Capita in China

Measure	Year	Municipality directly under central government	City at sub-provincial level	General provincial capital	General city at prefecture Level	City at county level	County seat	General town
Average investment per capita (in yuan)	2002	1895.92	1323.55	978.06	554.84	474.60	–	–
	2006	3438.31	2219.30	1615.53	1062.89	815.57	634.78	354.18
	2012	4594.25	5245.60	4204.57	3156.94	2074.21	2605.70	768.59
Relative level (1 for cities at the county level)	2002	3.99	2.79	2.06	1.17	1.00	–	–
	2006	4.22	2.72	1.98	1.30	1.00	0.78	0.43
	2012	2.22	2.53	2.03	1.52	1.00	1.26	0.37

Note The average investment per capita is calculated with urban population data for 2002 and urban-area population data for 2006 and 2012; for towns and county seats, this figure is calculated with populations in built-up areas and county seats respectively; the temporary population is always included

Source: *China Urban Construction Statistics Yearbook* (2002), *China Urban Construction Statistical Yearbook* (2006 and 2012) and *China Urban-Rural Construction Statistical Yearbook* (2006 and 2012)

institutional configurations since they are nearer to the centers of power, better satisfy conditions for pilot programs and are more likely to influence decision makers. In other words, institutional configurations in China currently favor administrative centers and cities at higher administrative levels. As an example, the Chinese government has tended to choose coastal cities at higher administrative levels to launch China's first economic and technological development zones, high-tech industry development zones, export processing zones, bonded zones and bonded port areas, before gradually setting up similar zones in other cities. The 10 "new areas" set up under the approval of the State Council, with the exception of the Zhoushan Islands New Area, are all in municipalities directly under the Central Government, cities at the sub-provincial level and provincial capitals. These include Shanghai, Tianjin, Chongqing, Lanzhou, Guangzhou, Xi'an, Guiyang, Anshun, Qingdao and Dalian. The government's 12 pilot zones for overall reform once again are all in municipalities directly under the Central Government, cities at the sub-provincial level and provincial capitals, with the exception of the plains in Shanxi and Heilongjiang Provinces and Yiwu City, Zhejiang Province. These include Shenzhen, Pudong New Area in

Table 6.6 Inter-city gaps in municipal facilities in China, 2012

Measure	Actual value						Relative level (1 for cities)	
	City	#Municipality directly under central government[a]	#City at Pre-fecture level[a]	#City at County level[a]	Town	#County seat	County seat	Town
Tap water penetration (%)	97.16	98.46	95.83	94.23	80.79	86.94	0.89	0.83
Gas penetration (%)	93.15	98.33	89.72	85.72	46.70	68.50	0.74	0.50
Sewage treatment rate (%)	87.30	88.19	85.50	81.34	<20[b]	75.24	0.86	<0.23
Average green space area per capita (m^2)	12.26	11.91	12.49	11.22	1.92	8.99	0.73	0.16
Green space percentage of built-up areas (%)	35.72	37.38	35.13	31.50	9.80	23.32	0.65	0.27
Green coverage of built-up areas (%)	39.59	40.58	38.96	35.96	13.70	27.74	0.70	0.35

Note [a]The arithmetic means are used for each city; [b]Data in 2010 is used
Source: *China Urban-Rural Construction Statistical Yearbook* (2012) and *China Urban Drainage and Wastewater Treatment Bulletin* (2006–2010) released by the MOHURD

Shanghai, Binhai New Area in Tianjin, Xiamen, Chongqing and Chengdu, as well as provincial capital-centric urban agglomerations and economic zones, such as the Wuhan-centric and Changsha-Zhuzhou-Xiangtan agglomerations and the Shenyang Economic Zone. As a further example, the Chinese government has approved the establishment of a pilot free trade zone in Shanghai. Building industrial parks, new areas and other functional zones stimulate the agglomeration of people, industries and factors of production as well as city size expansion. Also under the current administrative hierarchy, whether a city is allowed to have districts also depends on its administrative level. Cities at the prefecture or higher levels may have districts, whereas cities at the county level cannot have them. In recent years, upgrading counties or cities at the county level into districts has become the primary approach for municipalities directly under the Central Government, cities at the sub-provincial level and provincial capitals seeking to expand in size. This practice has furthered accelerated the size expansion of cities at higher administrative levels. In addition, China still lacks a sustainable local taxation system and a long-term mechanism for financing urban development. Local governments rely heavily on land-based revenues, namely, land transfer revenues and land mortgage, for urban development. Such institutional configurations favor larger cities at higher administrative levels over towns and smaller cities at lower administrative levels. This is because larger cities at higher administrative levels usually enjoy higher land and housing prices and land transfer revenues, plus easier financing through land mortgage, which are able to support large-scale urban development and operations. By comparison, towns and smaller cities at lower administrative levels are woefully devoid of similar resources.

Under these favorable power settings, resource allocation and institutional configurations, along with the effects of market forces, administrative centers at higher levels necessarily attract an increasing number of people, industries and factors of production from surrounding areas. Generally, the higher the administrative level of a city, the better able it is to attract resources and factors of production from surrounding areas. Such attractiveness is directly attributable to the city's administrative powers and government resource allocation. It is embodied not only in the agglomeration of people and industry, but more importantly in human and financial resources and even land quotas. Spatially, such attractiveness is manifested in two main aspects. First is the ability of administrative centers to attract resources from surrounding rural areas thanks to their power- and center-based advantages. Second is the ability of cities at higher administrative levels to attract resources from cities at lower administrative levels thanks to their administrative powers as well as better public facilities and services. A large number of local companies, for example, have moved their headquarters to cities at higher administrative levels such as Beijing and Shanghai, resulting in a continuous flow of resources from less developed regions to developed ones. This is also true in the case of rural migrant workers. In light of migration costs and ease of communication with family members, it seems that farmers should be more willing to find jobs in towns and small and medium-sized cities near home. In reality, however, farmers generally flock to large and very large cities, especially municipalities directly under the Central Government, cities at the sub-provincial level and provincial capitals, due to insufficient employment opportunities in towns and inter-city gaps in public services and expected income. In 2013, 63.9% of non-local rural migrant workers in China migrated to municipalities directly under the Central Government, provincial capitals and cities at the prefecture level, of which the former two combined account for 30.5%, according to an NBS survey (2014). The system under which counties are administered by cities is another typical example. This system is intended to eliminate the urban-rural divide and maximize the role of central cities in stimulating the growth of county economies. In reality, however, a lot of cities at the prefecture level more often take advantage of their central position and power to attract resources and factors of production from the administered counties. In addition to "pumping" financial and other resources from the administered counties, some cities at the prefecture level even employ administrative forces to "concentrate" land quotas granted to these counties into the former's central urban areas.

Evidence shows that the recent rapid expansion of larger Chinese cities, especially very large and ultra-large cites, is closely connected with the bias toward administrative centers in the process of urbanization. The government's bias toward cities at higher administrative levels, demonstrated in terms of power settings, resource allocation and institutional configurations, combined with market forces strongly fuel the rapid expansion of cities at higher administrative levels. To a certain extent, this bias is also manifested in a bias toward larger cities, as cities at higher administrative levels are usually larger in size. International studies show that bias toward larger or primary cities also exists outside China (Moomaw and Shatter 1996; Gugler 1982). Scholars Ades and Glaeser (1995) believe that, out of all the forces that drive agglomeration into cities, political forces are more important than economic factors. They further argue that a centralized system of government drives agglomeration into megalopolises. Under the effects of the government and market forces, therefore, a bias toward administrative centers in resource allocation will stimulate the agglomeration of people, industries and factors of production into these centers. The resulting economies of scale, in turn, will become an important basis for the government to decide on continued resource allocation. It is this mutually enhancing effect between the aforementioned bias and market forces that has caused the rapid expansion of larger cities and the failure of the Chinese government's size control policy in recent years.

6.4 Promoting Inter-city Equality in Opportunities and Rights/Benefits

To sum up, the size and growth rates of Chinese cities have much to do with their administrative level. Cities at higher administrative levels tend to be larger in size and grow faster than cities at lower administrative levels. These inter-city gaps result from the interplay between government bias toward administrative centers and market forces. The basis for such a gap is rooted mainly in the government's favoring of cities at higher administrative levels in terms of power settings, resource allocation and institutional configurations and the strong attraction of these cities due to the combined effects of the government and market forces. The bias toward administrative centers stimulates the development of cities at higher administrative levels but causes unfair competition between cities, distorts the resource allocation function of the market, and compels cities to compete for higher administrative levels. The excessive concentration of resources into cities at higher administrative levels, especially primary cities, will not only hurt economic growth (Tolley et al. 1979; Richardson 1987) but will also result in lower quality of life in non-primary cities (Henderson 2002). It is therefore necessary to eliminate both the bias toward administrative centers in resource allocation and the effects of administrative levels on the allocation of public resources. This will help promote inter-city equality in opportunities and rights/benefits, and create a fair and impartial external environment for the coordinated development of towns and cities of varying sizes. There are three main elements to accomplishing this goal.

1. **Control city size based on local carrying capacity**. China has long implemented a city size control policy based only upon population sizes. Towns and cities of varying sizes actually have different strengths and weaknesses, each performing different functions and roles in the urban system to form an organic whole. It is not always necessary to control the development of a city just because it is populous or encourage its development just because its population is small in size. Rather than the narrow focus on population, the main focus for size control policy should be on a city's growth potential and overall carrying capacity. Overall carrying capacity is defined in three main ways. First is the carrying capacity of local resources and the environment, which sets the upper limit for urban population size. Second is the capacity of urban infrastructure. Once an urban population exceeds this capacity, crowdedness or shortage of necessary resources lowers the quality of resident life. Third is the ability to accept new migrants, which depends on the availability of employment opportunities in the city. A shortage of industrial support would lead to insufficient employment opportunities. A large number of rural migrants would not be able to find work or would enter the informal sector for work. This would worsen urban poverty and result in social instability. For the future, China should base its city size control policy upon overall carrying capacity instead of population size. Size alone should not be a factor for government control, rather the focus should be on ensuring that the carrying capacity of local resources and the environment is not exceeded. The size and growth of a city must match its overall carrying capacity. As mentioned above, this study advocates human population size control for cities with urban-area populations of more than 4,000,000 people, not because of the very large size of these cities, but because most of these cities are near or have exceeded their resource and environmental carrying capacity.

2. **Maintain an environment for fair competition between cities**. Under the socialist market economy system, all cities are eligible for equal development opportunities regardless of their size or location. Inter-city fairness and equal opportunities are what China seeks to achieve. As described above, Chinese cities are categorized at different levels in the administrative hierarchy. It is difficult to eliminate these administrative levels altogether without the radical reformation of the city-based administrative division system. While it won't be realistic to eliminate this system altogether, it is important to restrict the expansion of these administrative levels, and remove administrative level as the basis for allocating public resources. Favoring larger cities at higher administrative levels is de facto discrimination against other cities. Administrative centers do not automatically have to develop into economic centers. It is necessary to properly handle the relationship between the government and the market during the process of urbanization under the principle of "allow the market to dominate, the government to offer guidance, farmers to decide for themselves and costs to be shared." Pervasive government problems such as excessive intervention and inaction must be resolved, and the market's decisive role in resource allocation must be enhanced. Global experience shows that market forces

will stimulate the concentration of people, industries and factors of production into larger cities, especially very large ones, in the early and intermediate stages of industrialization and urbanization. For towns and cities of varying sizes to develop in a coordinated manner, therefore, the government should conduct "reverse" regulation oriented toward eliminating market flaws and focus on supporting the development of towns and small/medium-sized cities. The government's regulatory policy will become redundant if it adopts a regulatory method that completely follows the market. The same goal can be achieved by improving the market mechanism. At present, the government should focus city size regulation on two things. First, it should decentralize certain functions away from very large and ultra-large cities that are close to exceeding or have already exceeded the local overall carrying capacity. These cities should be encouraged to finish transformation as soon as possible and improve its sustainable development. Second, the government should undertake measures to encourage the development of towns and small/medium-sized cities and enhance their industrial support and public services.

3. **Fuel the growth of towns by granting them more powers**. Granting more powers to towns will greatly facilitate their growth. In order to suit the needs of China's new urbanization model, it is necessary to implement a differentiation strategy so as to promote the growth of towns on a categorical basis. First, towns with population sizes that meet the standards for setting up a city should be upgraded to cities as early as possible. This can be achieved by upgrading the town or the entire county into a city. Moreover, pilot projects to upgrade towns into cities should be implemented in regions that meet certain requirements. Scientifically-sound standards for setting up cities should be built upon considerations such as urban-area population density, the rate of human agglomeration, municipal facilities, interregional differences and special needs, rather than administrative levels, economic aggregate and other indicators. Second, it is necessary to increase investment in public infrastructure in towns according to the relevant standards for small cities. This includes accelerating the construction of roads, public water/gas/heating supply, sewage and MSW treatment systems, building cultural/sports facilities in town areas, and improving local public services. This will gradually narrow the gap in public services between towns and cities. Third, given that it is difficult to upgrade towns into cities on a large scale—this will lead to an increase in administrative level and public expenditures—the focus of upgrading should be on the most economically developed towns with locational advantages and stronger abilities to accept migrants. Pilot reform programs to strengthen towns by granting them more administrative and socioeconomic management powers should be implemented. Such programs will encourage local governments to provide more industrial support and public services to towns, thereby increasing their growth potential.

References

Ades, A. F., & Glaeser, E. L. (1995). Trade and circuses: Explaining urban giants. *The Quarterly Journal of Economics, 110*(1), 195–227.

Cheng, Y. N. (2007). On city sizes and administrative levels in the Qing Dynasty. *Journal of Yangzhou University (Humanities and Social Sciences Edition), 3,* 124–128.

Deng, W. (2011). On state-owned economy, administrative levels and income gaps between Chinese cities. *Economic Science, 2,* 19–30.

Ding, J. H. (2011). On urban explosion and population control. *Forum for Advancement, 2,* 33–36.

Gao, P. Y. (1991). *Comparative research on urbanization in China and other countries*. Tianjin: Nankai University Press.

Gugler, J. (1982). Over urbanization reconsidered. *Economic Development and Cultural Change, 31*(1), 173–189.

He, X. G. (1991). On population expansion and size control in large Chinese cities. *Social Scientist, 2,* 42–50.

Henderson, V. (2002). Urban primacy, external costs, and quality of life. *Resource and Energy Economics, 24*(1), 95–106.

Hu, Z. L. (1986). On the ahead-of-the-times development of large cities and countermeasures. *Journal of Peking University (Philosophy & Social Sciences), 5,* 118–122.

Hu, Z. L., Liu, H. X., Ma, A. M., & Cai, Y. P. (1985). On the laws of the "ahead-of-the-times" population growth of large cities. *Social Survey and Research, 2,* 49–50.

Liu, Z. H., Mao, H., & Wang, L. Z. (2005). *Despotic powers in Chinese society*. Tianjin: Tianjin Ancient Books Publishing House.

Ma, Z. L. (1998). *Historical geography of Chinese cities*. Jinan: Shandong Education Press.

Moomaw, R. L., & Shatter, A. M. (1996). Urbanization and economic development: A bias toward large cities? *Journal of Urban Economics, 40*(1), 13–37.

National Bureau of Statistics. (2014). *Migrant workers monitoring survey report 2013*. July 25, 2014.

Rao, H. L. (1989). On cities' economies of scale. *Social Sciences in China, 4,* 3–18.

Richardson, H. W. (1987). The costs of urbanization: A four-country comparison. *Economic Development and Cultural Change, 35*(3), 561–580.

Tolley, G., Gardner, J., & Graves, P. (1979). *Urban growth policy in a market economy*. New York: Academic Press.

Wang, Y. (1985). On how to bring large cities' advantages into full play. *Social Survey and Research, 2.*

Wang, M. T. (2013). The sequence of administrative levels of cities and the urban & rural planning system. In *Collaborative planning in the era of cities – the collection of papers at the Annual National Planning Conference 2013*. November 2013.

Wang, Z. P. (2014). Control over population sizes of very large cities: A paradox. *China Development, 2,* 21–24.

Wei, H. K. (2010). On the scientifically-sound development of town economies in China. *Jianghai Academic Journal, 2,* 80–86.

Wei, H. K. (2014). Urbanization in China: Polarization and size pattern reconstruction. *China Industrial Economics, 3,* 18–30.

Wei, H. K., & Bai, M. (2008). On headquarters relocation of public companies in China: Status and characteristics. *China Industrial Economics, 9,* 13–24.

Wei, H., Wang, Y., & Bai, M. (2014). *The micro-analysis of regional economy in china: A perspective of firm relocation*. Singapore: World Scientific.

Yang, G. Q. (1998). On land supply and demand in Chinese cities: Status, reasons and countermeasures. *Urban Problems, 6,* 35–37.

Zhang, S. D. (2000). On urban governance: Forms and structures. In G. W. Skinner (Ed.), *The city in late imperial China* (pp. 98–104). Beijing: Zhonghua Book Company.

Zhao, G. (2006). *A collection of papers on the development of Chinese cities*. Beijing: New Star Press.
Zhou, Y. X. (1990). Review of guidelines for urban development in China. *Urban Economic Research, 1*, 1–16.
Zhou, C. S. (2001). *Cities in the Han Dynasty*. Beijing: People's Publishing House.

Chapter 7
Spatial Expansion Effect of Urbanization in China

With respect to urbanization in China, there is a very obvious phenomenon of uncontrolled and disorderly city sprawl. The amount of built-up areas and land used for urban development has long been increasing at a much faster rate than the urban population. This disorderly urban sprawl and a pattern of decentralized development have been widely criticized by scholars (Chris and Jay 2006; George 2001; Ewing 1994), especially since the emergence of urban growth management in the West in the 1960s, along with the rise of environmental protection movements and new theories and concepts regarding regionalism, new urbanism and smart growth (Li et al. 2005). This chapter presents empirical analysis of the spatial expansion effect of cities during urbanization in China, and presents the finding that future urbanization in China must be based on an approach that maximizes efficient and effective use of land and other resources.

7.1 Theoretical Analysis of the Spatial Expansion Effect of Cities

According to previous studies, the spatial expansion effect of cities includes economic, environmental and social effects (see Fig. 7.1) (Glaeser and Kahn 2004).

In terms of the economic effects, the spatial expansion of cities usually results in a decrease in the density of urban economic activities and, as is generally believed, in urban economies of agglomeration (Glaeser and Kahn 2004). Previous studies on density effect have shown this to be true (see Table 7.1). Ciccone and Hall (1996) were the first to study how the density of economic activities relates to productivity. After conducting empirical analysis of county-level data in the United States, they found that average productivity would increase by 6% for every 100% increase in the density of economic activities at the county level. Later on, Ciccone (2002) researched county-level data in European countries and found that the elasticity

H. Wei, *Urbanization in China*, Research Series on the Chinese Dream and China's Development Path, https://doi.org/10.1007/978-981-13-1408-7_7

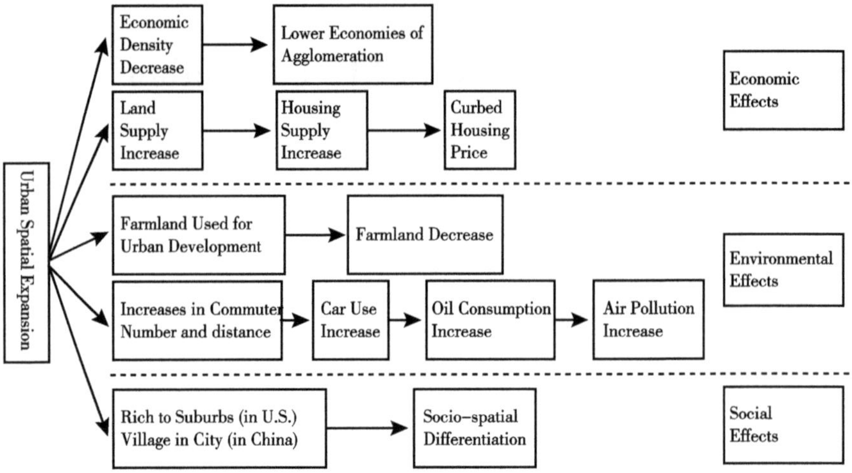

Fig. 7.1 Mechanism for the spatial expansion effect of cities

of productivity relative to economic density was 4.5%. A study by Combs found that productivity measured by wages/salaries would increase by 3% for every 100% increase in employment density. In China, Fan (2006) analyzed 2004 data from Chinese cities at the prefecture or higher levels and found that the elasticity coefficient of non-agricultural productivity relative to non-agricultural employment density was 8.8%. Chen et al. (2008) analyzed data from Beijing-administered districts and counties and found that the elasticity of productivity relative to output per unit area and employment was 11.8 and 16.2% respectively. Liu (2009) studied 2003–2006 panel data from Chinese cities and found out that urban productivity would increase by 1.7% for every 10% increase in employment density.

Meanwhile, the spatial expansion of cities will lead to increases in urban land supply and the corresponding area of land used for housing development, which is helpful in preventing the fast rise of urban housing prices and increasing the average housing area per capita. Nonetheless, if spatial expansion does not match a city's ability to accept new migrants, then problems such as empty towns will likely follow (Wei and Su 2013). This phenomenon is exemplified by the cities of Ordos and Tieling in China.

With respect to environmental effects, the spatial expansion of cities affects the environment in two main ways. The most immediate effect is the decrease in farmland and woodland area and quality as a result of urban development. In addition, spatial expansion tends to increase car use, reflected in both a growing number of cars on the road and increasing commuting distances. This in turn leads to increased fuel consumption and pollution. International city data shows that the wider the urban sprawl, the greater the number of car users (Newman and Jeffrey 1998; Kahn 2000). Nonetheless, Glaeser pointed out with U.S. data that urban sprawl will not produce a significant environmental impact thanks to the increasing use of eco-friendly vehi-

Table 7.1 Previous studies on urban density effect

	Sample	Dependent variable	Independent variable	Elasticity coefficient	Instrumental variable
Ciccone and Hall (1996)	U.S. State level data in 1988	Productivity	Employment density	6%	Overall index[a]
Ciccone (2002)	European countries	Productivity	Employment density	4.5%	Land area
Combes et al. (2008)	France (1976–1998)	Wages/salaries	Employment density	3%	Population density in 1936
Fan (2006)	Chinese city data in 2004	Productivity wages/salaries	Employment density	8.8%	Urban land area
Chen et al. (2008)	Beijing district/county data in 2004	Productivity	Output density employment density	11. %, 16. 2%	Density in 2000
Liu (2009)	Panel data on Chinese cities (2003–2006)	Productivity	Employment density	17%	Railway status in 1933

Note [a]The overall index includes railway status in 1860 as well as population status and distances to the east coast in 1850 and 1880

cles. In addition, some scholars emphasize that urban sprawl may damage wetlands, reduce green spaces and reduce the efficiency of land use and other resources (Wang et al. 2014).

With respect to social effects, the spatial expansion of cities tends to lead to a dual structure within cities, according to the history of global urban expansion. This structure is manifested in two primary ways. First is the division between the rich and poor, apartheid or social isolation. In the United States, this is exemplified by rich people moving to suburbs. The rich are more willing to move to city outskirts than the poor because they travel mainly by car and can afford larger houses. It is also believed by some that the social division is much less obvious in suburbs than in urban areas (Cutler et al. 1999) and that suburban residents are less likely to be members of social organizations (Putnam 1999), and so forth. The second manifestation is the appearance of a large number of villages in the city and shanty towns, which is more common in China than the United States. In either form, spatial expansion tends to form a dual spatial structure and dual social structure within a city. This process also leads to social problems such as disputes arising from land expropriation and housing demolition.

Table 7.2 lists the strengths and weaknesses of urban sprawl and compact cities. In combination with the situation in China, the following section analyzes how the spatial expansion of cities affects economies of agglomeration, fuels car consumption and carbon emissions, and causes dual structures within cities.

Table 7.2 Urban sprawl versus compact cities

	Advantages	Disadvantages	Examples
Sprawl	✓ High livability, including larger houses, more comfortable surroundings, and so forth ✓ Curbed housing prices	• Decreases in farmland size and quality • Oil consumption and air pollution by cars • Lower economies of agglomeration • Possible empty towns • Faster spatial and social differentiation	The United States Australia Empty towns in China
Compact	✓ Higher economies of agglomeration ✓ More efficient/effective land use ✓ Lower fuel consumption and exhaust emissions	• Crowdedness • High housing prices	Europe Japan

7.2 Correlation Between Spatial Expansion of Cities and Productivity

The density of urban economic activities affects urban productivity by affecting economies of agglomeration. With this in mind, we developed an urban production function including economies of agglomeration to study how the density of urban economic activities affects urban productivity:

$$\mathrm{Y/L} = \mathrm{A}e^{f(AC)}(K/L)^{\alpha}H^{\beta} \tag{7.1}$$

Take a logarithm on both sides of this equation:

$$ln\left(\frac{Y}{L}\right) = cons + f(AC) + \alpha ln\left(\frac{K}{L}\right) + \beta lnH \tag{7.2}$$

where Y is total output value, K is capital, L is labor, H is human capital (used to represent the quality of labor). Y/L denotes productivity, K/L denotes the average labor-capital ratio and A denotes all determinants of total-factor productivity (TFP) other than urban economies of agglomeration. In addition, AC denotes the city's agglomerative characters, or the effects of economies of agglomeration on productivity. If $f(AC)$ is positive, namely, greater than 1, economies of agglomeration are contributive to productivity. According to Ciccone and Hall (1996), the elasticity coefficient in $f(AC)$ is the net economies of agglomeration including both economies of agglomeration and crowdedness effect.

According to previous studies, the city's agglomerative characters mainly include the size, density and specialization (structure) of urban agglomerative economic activities (Alonso 1970; Sveikauskas 1975; Segal 1976; Ciccone and Hall 1996; Ciccone 2002; Henderson et al. 1995; Henderson 1997; and so forth.). Since we focus on density effects herein, we take size effect and specialization's structure effects as important control variables:

$$f(AC) = f(size) + f(density) + f(specialization) \tag{7.3}$$

In considering the presence of optimal urban population density, let:

$$f(density) = a\, density + b\, density^2 \tag{7.4}$$

The panel data model ultimately used can therefore be expressed as:

$$\begin{aligned} ln\left(\frac{Y_{i,t}}{L_{i,t}}\right) \textbf{ or } ln\, w_{i,t} &= cons_{i,t} + a\, density_{i,t} + b\, density_{i,t}^2 + (a_1 size_{i,t} + b_1 size_{i,t}^2) \\ &+ (a_2 stucture_{i,t} + b_2 stucture_{i,t}^2) + \alpha ln\left(\frac{K_{i,t}}{L_{i,t}}\right) + \sum \beta_j ln H_{j,i,t} \\ &+ \gamma_1\, region_{i,t} + \gamma_2\, citylevel_{i,t} + \mu_{i,t} \end{aligned} \tag{7.5}$$

$ln(\frac{Y_{i,t}}{L_{i,t}})$where and $lnw_{i,t}$ denote the logarithms of the average output value and wage/salary per capita as dependent variables respectively. The linear and quadratic terms of the density measure, that is $(a\, density + b\, density^2)$ are introduced to study density effect and the optimal urban population density. $(a_1 size_{i,t} + b_1 size_{i,t}^2)$ and $(a_2 structure_{i,t} + b_2 structure_{i,t}^2)$ denote the introduction of the linear and quadratic terms of the size and specialization measures to control size and structure effects respectively. $ln(\frac{K_{i,t}}{L_{i,t}})$ is the logarithm of the average labor-capital ratio; $\sum \beta_j ln H_{j,i,t}$ denotes the number of teachers at each level used to replace human capital; $region_{i,t}$ and $citylevel_{i,t}$ denote control region and city level as two dummy variables respectively. Subscripts i and t denote city and timeframe respectively. Table 7.3 explains these variables and measures.

Dependent variables. The average output value per capita (total output value/the total number of employees, denoted as "*Y/L*") and average wage/salary (*w*) are taken as the measures of productivity.

Density variable. To address the endogeneity of employment density, urban passenger transport density is taken as the main variable for measuring the density of economic activities, primarily because the total passenger transport volume is not directly connected with productivity. Moreover, the total passenger transport volume in a city does directly reflect the economic activity within a city. Accordingly, its density is able to reflect economic activity per unit area, which is directly connected with the density of economic activities. Population density (*P*/*Area*) and passenger transport density (*T*/*Area*) are taken as the main measures for the density of economic activities. Population density equals total urban-area population (*P*) divided by built-up area size (*Area*). The average total passenger transport volume per city

Table 7.3 Variables and measures

		Measure	Description	Unit
Dependent variable	Productivity	Average output value per capita (Y/L)	Total non-agricultural output value/total number of employees in non-agricultural organizations	Yuan
		Average wage/salary (w)	Average wage/salary per employee	Yuan
Main explanatory variable	Density	Population density ($P/Area$)	Total urban-area population/built-up area size	1000 people/km^2
		Passenger transport density ($T/Area$)	(Total annual passenger transport volume by bus/built-up area size)/365 days	1000 people/day/km^2
Control variables	Size	Population size (P)	Total urban-area population size	1000 people
	Specialization	Relative specialization (RS)		–
	Capital	Average labor-capital ratio (K/L)	Total investment in fixed assets/total number of employees in non-agricultural organizations	Yuan
	Human capital	The number of university teachers (gx) The number of secondary school teachers (zx) The number of elementary school teachers (xx)	The numbers of various teachers	People
	Dummy variables	Region (*region*)	East, Central, West and Northeast	–
		City level (*citylevel*)	Cities at the prefecture or higher levels	–

(T) is expressed as the daily average passenger transport volume per city, which equals (the annual total passenger transport volume/built-up area size)/365 days.

Control variables. To control the effects of size and specialization on productivity, size and specialization are introduced as two control variables, which are expressed as total urban-area population and relative specialization respectively. The relative specialization coefficient proposed by Duranton and Puga (2005) is used for relative structure. S_{ij} is defined as the percentage of Industry j in all employees in City i, namely, $S_{ij} = L_{ij}/L_i$, where L_{ij} denotes the number of employees in Industry j in City$_i$ and L_i denotes the total number of employees in City$_i$; S_j is the percentage of Industry$_j$ in the total number of employees across China. The relative specialization index of a city is expressed as: $\mathrm{RS}_i = \frac{1}{2}\sum_j |S_{ij} - S_j|$. Given that the size and structure effects may also be characterized by an upside-down U, the linear and quadratic terms of both variables are introduced directly as control variables.

The average labor-capital ratio (K/L) is expressed as total investment in fixed assets/the total number of employees in non-agricultural organizations. The human capital measure is replaced with the number of teachers, which is controlled by the numbers of university teachers (gx), secondary school teachers (zx) and elementary school teachers (xx). Two dummy variables—cities at the prefecture or higher levels and region—are set to control the effects of cities' administrative levels and regions respectively.

Econometric analysis is made with data from 284 Chinese cities at the prefecture or higher levels from 2006 to 2010 mainly taken from *China City Statistical Yearbook* (2007–2011) and *China Urban Construction Statistical Yearbook* (2006–2010). Of these cities, Lhasa, Zhongwei and Longnan are not considered due to incomplete data.

Table 7.4 offers the result of density effect test.[1] With respect to the capital and human capital variables, the average labor-capital ratios in the models are all around 0.5 and are significant at level 1%, which are consistent with their economic meanings, when the average output value per capita is a dependent variable. In addition, the average labor-capital ratios in the model are all around 0.05 and are significant at level 1% when the average wage/salary is a dependent variable. The latter case shows that the average labor-capital ratio has weaker effects on wages/salaries, which happens to reflect the economic meaning of wages/salaries as a measure for net productivity and supports the reasonableness of the fact that, in most previous studies, models with wages/salaries as a dependent variable did not include the variable of capital. Models 1 and 2 are consistent with previous studies in terms of the test result concerning positive elasticity coefficient, according to the result of density effect test. In Models 3 through 6, the coefficients of linear terms of passenger transport density in two models are all positive, those of its quadratic terms are all negative, and they are all significant at level 1%, with a significant upside-down U-shaped relationship with both the average output value per capita and wages/salaries. We can thus see that density effect is significantly characterized by an upside-down U

[1]The result of independent test on density effect is omitted herein. For more information, please refer to Wei and Su (2013).

Table 7.4 Density effect test result

	Model 1	Model 2	Model 3	Model 4	Model 5	Model 6
	Independent variable					
Dependent variable	Ln(*Y/L*)	Ln(*w*)	Ln(*Y/L*)	Ln(*w*)	ln(*Y/L*)	Ln(*w*)
C	4.6383[b] (20.75)	9.0144[b] (62.31)	5.1896[b] (23.45)	9.1790[b] (62.30)	5.6449[b] (22.67)	9.4431[b] (60.31)
T/Area	0.2110[b] (5.27)	0.1327[b] (5.11)	0.3466[b] (4.67)	0.3297[b] (6.68)	0.3366[b] (4.53)	0.3022[b] (6.45)
$(T/Area)^2$			−0.1579[b] (−3.32)	−0.1610[b] (−5.09)	−0.1422[b] (−2.98)	−0.1313[b] (−4.35)
Ln(*K/L*)	0.5285[b] (32.45)	0.0689[b] (6.53)	0.4944[b] (30.63)	0.0589[b] (5.48)	0.4897[b] (30.59)	0.0647[b] (6.36)
P					0.0008[b] (3.92)	0.0009[b] (7.45)
P^2					−3.20e−07[b] (−3.28)	−2.95e−07[b] (−4.80)
RS					0.7779[a] (2.21)	0.4125[b] (7.40)
RS^2					−1.6795[b] (−3.13)	
The numbers of various teachers	Controlled	Controlled	Controlled	Controlled	Controlled	Controlled
Region	Controlled	Controlled	Controlled	Controlled	Controlled	Controlled
Cities at the prefecture or higher levels	Controlled	Controlled	Controlled	Controlled	Controlled	Controlled
Total R^2	0.6377	0.3271	0.5618	0.3813	0.5742	0.4502
F-statistic	195.51	85.45	162.81	78.28	121.92	79.80
Hausman H-value	18.11	470.65	25.15	473.70	30.35	486.48
Observed value	1316	1316	1285	1285	1285	1285

Note Parenthesized figures are t-test values; [a] and [b] mean that they are significant at levels 1 and 5% respectively. This is also true with the following tables shape regardless of dependent and control variables. With respect to population size and relative specialization as two control variables, the coefficients of linear terms of both variables in Model 5 are all positive, those of their quadratic terms are all negative, and they are all significant at a higher level when the average output value per capita is a dependent variable. Both size and structure effects are significantly characterized by an upside-down U shape. In Model 6, size effect is still significantly characterized by an upside-down U shape when wages/salaries is a dependent variable, whereas structure effect only shows a positive correlation (since there is no correlation when quadratic terms are introduced, only the result of models with linear terms introduced is listed herein).

Since the total passenger transport volume is unable to directly reflect population density, the same panel data is used to make a regression analysis of urban-area population density and passenger transport density. Fixed effect is selected for the Hausman test. Both constant and coefficient terms are all significant at level 1%, with the result as follows:

$$\begin{array}{lcccc} \text{Urban-area population density} = & 0.8337 & + 0.4062 \times \text{passenger transport density} & + \mu \\ \text{t-value} & (46.85) & (12.62) & (7.6) \end{array}$$

Coefficients in Models 3 through 6 are used to calculate the optimal passenger transport density in each model, before the optimal urban population density is estimated according to the aforementioned equation. The estimates are shown in Table 7.5. The result shows that the optimal urban population density is around 13,000 people/km^2 if it is calculated as total urban-area population/built-up area size.

Table 7.5 Estimates of optimal urban population density

	Model 3	Model 4	Model 5	Model 6
Coefficients of first-order terms	0.3466	0.3297	0.3366	0.3022
Coefficients of second-order terms	−0.1579	−0.1610	−0.1422	−0.1313
Optimal value of passenger transport density (in 1000 people/km^2)	10.975	10.239	11.835	11.508
Estimated optimal urban population density (in 1000 people/km^2)	12.795	12.496	13.145	13.012

7.3 Correlation Between Spatial Expansion of Cities, Car Increase and Carbon Emissions

The environmental impact of the spatial expansion of cities includes decreases in farmland size and quality and increases in fuel consumption and the emissions of carbon and pollutants such as SO_2 due to the increased use and number of cars. Since this study has already analyzed how the spatial expansion of Chinese cities decreases high-quality farmland,[2] this section focuses its analysis on how expansion affects the number of vehicles and thus fuels consumption and car emissions.[3]

First, based on theoretical analysis, this study assumes that car demand is positively correlated with built-up area size. Figure 7.2 illustrates this simple correlation using Chinese city data. Car demand is also affected by total human population, population density, and level of economic development and public transport. Specifically, higher car demand will result from larger built-up area size, stronger demand for transport, larger total human population, larger urban sprawl with lower urban population density or higher level of economic development. A well-developed public transport system will somewhat reduce car demand. Second, fuel consumption is positively correlated with car count, which is illustrated by Fig. 7.3 using Chinese data, and is closely connected with the average commuting distance, which depends on built-up area size. In a strict sense, the average commuting distance can

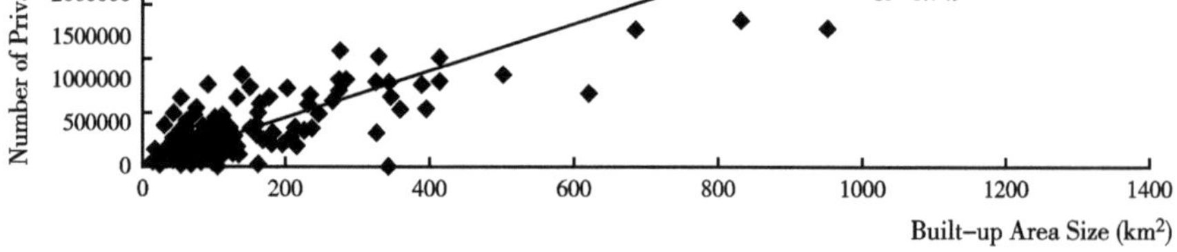

Fig. 7.2 Correlation between built-up area size and number of privately owned automobiles in Chinese cities, 2010. Note: Data from Chinese cities at the prefecture or higher levels is used; car count is expressed as civilian vehicle population. Source: *China Urban Construction Statistical Yearbook* (2010) for built-up area size data, and *China Statistical Yearbook for Regional Economy* (2011) for civilian vehicle count in Chinese cities

[2]Rapid urban sprawl in China has consumed a large amount of farmland and has led to decreases in farmland size and quality.

[3]Only carbon emissions are included herein since SO_2 emissions relate to both the sulfur content in fuel and other energy mixes. Namely, they may drop with a decrease in the former and changes in the latter.

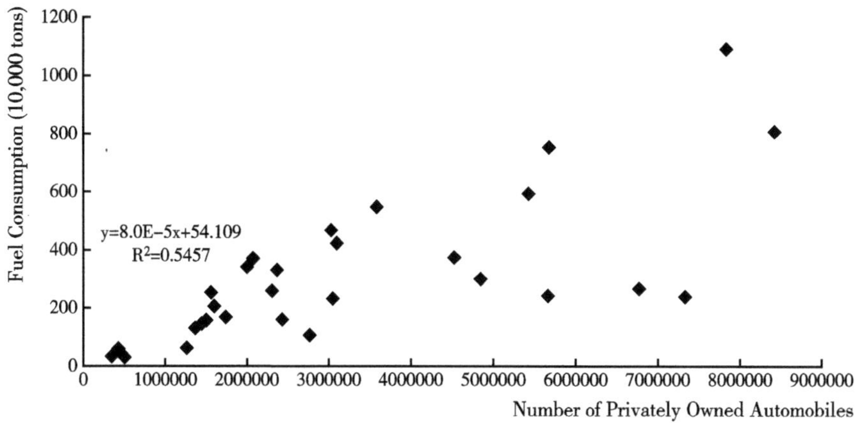

Fig. 7.3 Correlation between number of privately owned automobiles and fuel consumption by province, 2010. Source: *China Energy Statistical Yearbook* (2011) for fuel consumption in provinces and *China Statistical Yearbook for Regional Economy* (2011) for civilian vehicle count

be expressed as $F = Cxr$, where F, C and r denote fuel consumption, car count and the average commuting distance respectively. According to Bettencourt (2013), this expression can be simplified as $r = \alpha A^{1/2}$, where A is built-up area size. As can be seen from this data, fuel consumption is correlated with both car count and built-up area size.

To further test this correlation and obtain relevant coefficients, the following econometric model was formulated according to theoretical analysis and data availability, with relevant variables controlled (the level of the public transport system as a variable is not controlled herein due to the unavailability of relevant measures and data):

$$C_{i,t} = aA_{i,t} + b\frac{P_{i,t}}{A_{i,t}} + cP_{i,t} + dE_{i,t} + \epsilon_{i,t} \tag{7.7}$$

where C denotes total car count, which is expressed as civilian vehicle count given data availability, P denotes total urban-area population, A denotes built-up area size, P/A denotes human population density, E denotes the average disposable income per urban resident, ε is the error term, and i and t denote city and timeframe respectively. 2006–2010 panel data from *China City Statistical Yearbook*, *China Urban Construction Statistical Yearbook* and *China Statistical Yearbook for Regional Economy* is used. The test result is shown in Table 7.6. This result indicates that the coefficient is 2122.65 and significantly positively correlated at level 1% when the relationship between built-up area size and car population is separately tested; the coefficient of the effects of built-up area size drops after control variables are introduced respectively—in Models 2 through 4, car count increases by about 800–1000 units for every 1-km^2 increase in built-up area size, and they are significantly positively correlated at level 1%. Under Models 3 and 4, car count increases by about 31,000–34,000

Table 7.6 Test on the correlation between urban expansion and car count

	Model 1	Model 2	Model 3	Model 4
C	17,103.76[a] (2.89)	31,492.46[a] (5.69)	64,956.10[a] (5.51)	38,946.29[a] (2.54)
Area	2122.65[a] (62.87)	1124.47[a] (15.80)	924.32[a] (9.79)	883.23[a] (9.25)
P		813.11[a] (15.60)	982.75[a] (13.27)	989.85[a] (13.38)
P/Area			−31,058.19[a] (−3.22)	−34,223.23[a] (−3.52)
E				2.30[a] (2.63)
Total R^2	0.7320	0.7660	0.7682	0.7735
F-statistic	3952.22	2444.11	1643.94	1239.01
Hausman H value	40.87	44.84	119.88	79.82
Observed value	1381	1380	1380	1379

units for every 10,000-people/ km^2 decrease in urban-area population density; under Model 4, resident income is positively correlated with car count.

Given the limited availability of data due to the lack of complete statistics about fuel consumption in Chinese cities, the ratio of fuel consumption to private passenger vehicle count is used to compute the annual average fuel consumption per vehicle. According to Fig. 7.4, this average consumption is estimated to be 218 kg in 2012 and has decreased in more recent years. On one hand, this downward trend relates to more stringent emissions standards for cars[4] and the increasing use of eco-friendly vehicles, which supports the view of Glaeser that the increasing use of eco-friendly vehicles can effectively reduce the environmental impact of urban sprawl. On the other hand, average car use has also been decreasing despite the growing number of private passenger vehicles due to license plate number-based limitations that have been implemented in cities across China, such as Beijing (since 2008), Nanchang (since 2009), Lanzhou (since 2010), Changchun (since 2010), Hangzhou (since 2011) and Chengdu (since 2011). By excluding the effects of decreasing average motor vehicle usage on fuel consumption per vehicle, therefore, we take the annual average fuel consumption before 2008 as the average fuel consumption per vehicle, which equals 330 kg/vehicle/year in the following calculation.

With parameter estimates in Table 7.6 and coefficients in Fig. 7.4, further analysis can be made on how growing built-up area size affects car count, fuel consumption and carbon emissions (see Table 7.7). The result shows that the number of cars increases by about 1000 units for every 1-km^2 increase in built-up area size if only separate effects are considered. With annual average fuel consumption per car of

[4]China released emissions standards for light vehicles in July 1999 and revised it in 2001. It successively implemented emissions standards equivalent to Euro I, II, III and IV emissions standards in 2001, 2004, 2007 and 2010 respectively.

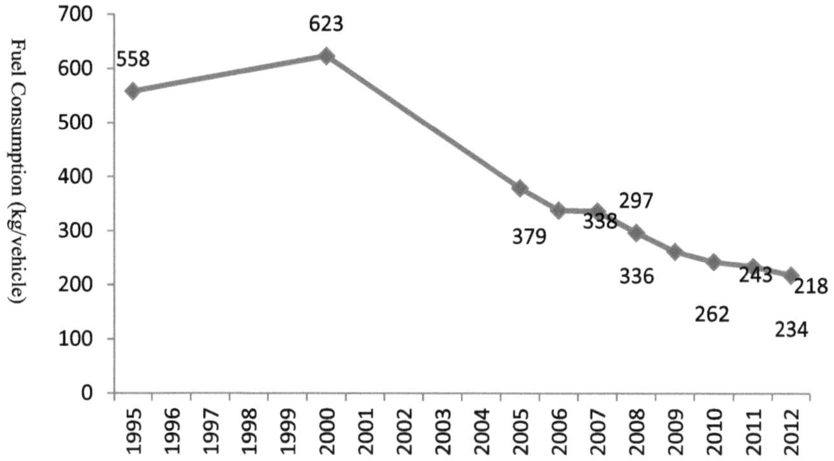

Fig. 7.4 Changes in average fuel consumption per vehicle. Source: *China Energy Statistical Yearbook* (2013) and *China Statistics Yearbook* (2013)

Table 7.7 Separate and combined effects of growing built-up area size on resources and environment

	Separate	Combined
Built-up area size	+1 km^2	+1 km^2
Car count	+1000 units	+2000 units
Fuel consumption coefficient	330 kg/vehicle/year	330 kg/vehicle/year
Fuel consumption	+330 tons	+660 tons
Standard amount conversion factor	1.4717	1.4717
Carbon emissions coefficient	0.5532	0.5532
Car emissions	+270 tons	+540 tons

Note Separate effects refer only to the correlation between built-up area size and car count without considering changes in other factors such as human population, density and income. Combined effects refer to how built-up area size, human population, density and other factors combined affect car count. For both standard amount conversion factor and carbon emissions coefficient, this study refers to relevant results of the Sustainable Development Strategy Study Group of the Chinese Academy of Sciences (2009)

about 330 kg, fuel consumption and carbon emissions will increase by about 330 and 270 tons respectively. If combined effects are considered, car count, fuel consumption and carbon emissions will increase by about 2000 units, 660 tons and 540 tons respectively.

7.4 Correlation Between Spatial Expansion of Cities and Dual Urban Structures

An important characteristic of urbanization in China is the lack of shanty towns that exist in other developing countries. However, there is now a growing residential space differentiation and emerging dual spatial structure in Chinese cities resulting from China's fast urbanization, especially the rapid spatial expansion of cities. This dual spatial structure is characterized by the dramatic difference between the forest of skyscrapers in certain areas compared with the sea of hovels in other areas. This spatial differentiation has become a tough challenge for urban development and management in some Chinese cities, especially old industrial centers and larger cities. In addition to this dual spatial structure, there is a growing dual social structure in Chinese cities. Rural migrants aside, this differentiation is occurring between poor urban residents living in shanty towns/villages in the city and rich ones living in affluent areas, an obvious phenomenon during urban sprawl. Such dual structure is unique to Chinese cities.

Municipal governments or developers tend to prefer acquiring low cost farmland, but avoid acquiring land in larger villages, given the costs of housing demolition and villager resettlement. Consequently, a large number of villages have been retained during urban expansion and are referred to as "villages in the city". The issue of "villages in the city" is especially severe in such cities as Beijing, Tianjin, Chongqing, Shanghai, Wuhan, Guangzhou and Shenzhen. There are three main problems with the status of villages in the city. First, villagers usually build houses without permission due to lack of strong authority regarding land management and urban planning. A lot of villages in the city have building densities over 70% or even 90%. Consequently, phenomena known as "shake-hand buildings", "face-to-face buildings" and "a thin strip of sky" is all too common. Second, such low-quality buildings provide non-locals with low-rent dwellings, resulting in a high percentage of non-locals in villages in the city, high mobility, diverse resident mix and security challenges. Third, villages in the city generally cannot benefit from municipal governments' infrastructure development programs since they are not covered by urban plans. Moreover, villagers or village committees are unable to build infrastructure on a large scale due to problems with land ownership, planning power and funding. Consequently, transport, water/power supply, sewage drainage and MSW treatment infrastructure in villages in the city are sub-par.

In addition, a growing number of shanty towns have emerged amid urban expansion. Houses in shanty towns provide the worst residential conditions in China. These include houses built by employers for factory workers or poor people before the People's Republic of China was founded in October 1949, houses built in or around factories, mines and forestation/farming zones since October 1949, and makeshift dwellings built by residents with or without government approval. Houses in these shanty towns are characterized by low resistance to disasters, crude and aging structures and poor functionality, combined with a bad living environment, a lack of space for public events and lack of supportive facilities. Moreover, households living in shanty towns are generally low-income or lower-middle-income households that cannot afford housing.[5] Shanty towns in China are similar to slums in other developing countries in terms of residential conditions and resident income. The presence of shanty towns has produced significant adverse effects on the city's appearance while resulting in the isolation of shanty town residents from those that reside in surrounding communities. The growing urban social differentiation in China is a reflection of this dual spatial structure. Between 2008 and 2012, 12.60 million houses in shanty towns across China were rebuilt, but there remain over 12.00 million low-income or lower-middle-income urban households living in shanty towns, according to the *Opinion of the State Council on Accelerating Shanty Town Rebuilding* (GF [2013] No. 25). This policy document sets a goal of rebuilding 10.00 million more houses in shanty towns in between 2013 and 2017. Unfortunately, however, the government faces great difficulty in rebuilding shanty towns, due to the costs and complexity of housing demolition and resettlement, along with poor existing infrastructure.

Besides the increasingly obvious dual spatial structure, a dual social structure is growing in Chinese cities. This dual social structure is manifested in two main aspects. First, there is the severe issue of a dual structure between rural migrants and native urban residents. According to 2012 data (see Fig. 7.5), the average urbanization rate based on permanent population in 35 very large and ultra-large cities[6] was 71.7%, yet the urbanization rate based on population possessing a local urban *hukou* was a mere 38.8%. There is a difference of 32.9 percentage points between these two indicators, compared with an average difference of 17.3 percentage points nationwide. In other words, there is a gap of 15.6 percentage points between larger cities and the nationwide average, which makes the dual structure more acute in very large and ultra-large cities as the main receivers of migrants. Of these cities, Shenzhen has the biggest difference between its permanent population and population with an urban hukou. In 2012, for example, Shenzhen had an urban permanent population of 10.5474 million people, but only 2.992 million people with local urban *hukou*. Shenzhen would have a 100% urbanization rate based on its permanent population, but only a 28.4% urbanization rate based on its population with local urban *hukou*, which is a gap of 71.6 percentage points. Rural migrants into Shenzhen are usually less educated and are far behind native urban residents in terms of political rights, employment, social security and access to public services.

[5]By the end of 2008, there were 11.48 million households living in shanty towns across China, including 7.44 million households in cities (64.80%), 2.38 million households in state-owned factories and mines (20.73%) and 1.66 million households in forestation/farming zones (14.46%); of these, 6.81 million households were low-income and cannot afford housing and 4.67 million households were lower-middle-income and cannot afford housing, according to a survey. There remain about 12.00 million households living in shanty towns due to adjustments to China's shanty town standards. The percentages of various shanty towns remain stable.

[6]Given that data for Lhasa is unavailable, these 35 cities include municipalities directly under the Central Government, provincial/regional capitals except for Lhasa, and cities specifically designated in the state plan, each with an urban-area population of over 1,000,000 people. This is also true with the following sections.

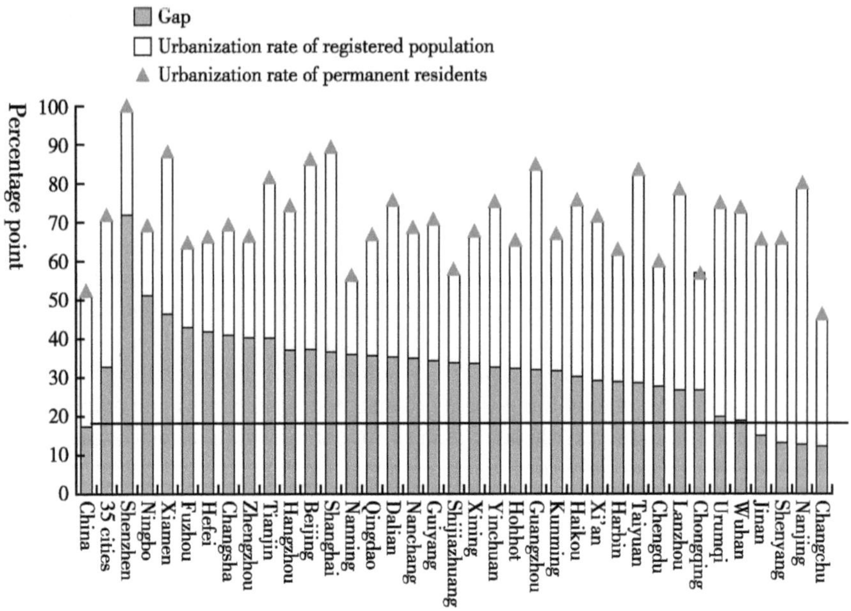

Fig. 7.5 Urbanization rates of registered population and permanent residents in 35 mega cities. Note: These 35 cities are sorted by size. Relevant data slightly differs from real values given the mobility of people with urban *hukou*. Source: Statistical bulletins 2012 released by cities and other relevant data

The second manifestation of the dual structure in cities is the wide gap in resident income and urban poverty. The population of poor people in Chinese cities has been growing in recent years. In other words, urban poverty is worsening. Between 1981 and 1990, China had an urban poor population of between 0.5 and 3.90 million people, with an average poverty rate of 0.5% (World Bank 1993). This is in stark contrast to China's current urban poor population[7] of between 20 and 30 million people (Wei and Wu 2009). A part of this problem is rural migrant workers' incomes. Despite continuously increasing in recent years, the average income of rural migrant workers remains under 60% of that of urban employees. Increasing from 1417 yuan in 2009 to 2290 yuan in 2012, the monthly average income of non-local rural migrant workers is still equivalent to only 58.83% of the monthly average income of urban employees, compared with 52.47% in 2009. Between 2011 and 2012, there was no rise, with the percentage remaining stagnant at about 58.8% (see Fig. 7.6).

[7]Chinese cities' poor population mainly includes the unemployed, people suffering from serious or long-term diseases and/or disabilities, low-income employees and other disadvantaged groups.

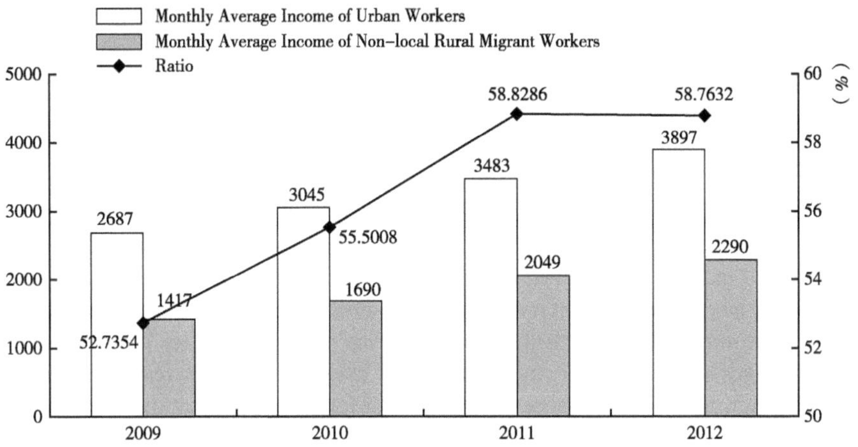

Fig. 7.6 Growth of non-local rural migrant workers' monthly average income, 2009–2012. *Source Migrant Workers Monitoring Survey Report* in relevant years and *China Statistics Yearbook*

7.5 Correlation Between Optimal Urban Population Density and Potential for Land Conservation

Analysis shows that the rapid spatial expansion of Chinese cities has adversely affected economic efficiency, resources and the environment, urban spaces and social structure. Future urbanization in China should focus on higher quality urbanization and resource efficiency. At the Central Economic Work Conference in 2013, the Chinese government emphasized that the concept and principles of eco-civilization should be fully embodied throughout urbanization and that future urbanization should be characterized by high resource efficiency, intelligence, green development and low carbon emissions. Resource efficiency refers to the efficient and effective utilization of resources, especially land, during urbanization. In this regard, emphasis is placed on unleashing the potential of urban land and conserving land resources through a pattern of urban development characterized by compactness, concentration and high efficiency. A resource-efficient approach to urbanization is necessary and feasible, according to the status of land use in Chinese towns and cities.

Land and population urbanization should also be conducted in accordance with optimal urban population density. Built-up area size expansion should adapt to urban population growth in line with optimal urban population density. On one hand, a city's ability to accept new migrants should support expansion of built-up areas so as to avoid the phenomenon of ghost or empty towns, which would leave land idle or inefficiently used. On the other hand, urban population growth should also adapt to built-up area size under the constraints of optimal urban population density. An excessively high population density will lead to crowdedness and lower environmental quality and livability, worsening urban diseases. A more pervasive and severe problem for most cities is not crowdedness, but land idleness and inefficient use arising from poorly controlled urban sprawl in earlier stages of urbanization. This exacerbates the already significant lack of land supply for future urbanization. Population urbanization in China will continue at a fast rate in the next two decades, but land supply in towns and cities is already at a bottleneck. In the aftermath of rapid urban sprawl in earlier stages of urbanization, land supply in China will be a limiting factor for China's further urbanization, industrialization and modernization, according to a study (Tu et al. 2013). Local governments have actually begun using various methods to increase land supply, such as reclaiming land from hills or seas, fully using land for rural development, redeveloping so-called "second-hand land" and using land in an efficient and effective manner. These last two methods are the most environmentally friendly approaches. Given poorly controlled urban sprawl in earlier stages and increasingly tight land supply in towns and cities, resource-efficient urbanization is therefore a necessary approach for new urbanization with Chinese characteristics in the future.

Based on the optimal urban population density value, this study analyzes the potential of land conservation in Chinese cities using 2011 data on total urban-area population and built-up area size in 657 cities. The result may serve as a reference for determining a scientifically-sound pattern of land use in future urbanization. Analysis of optimal urban population density indicates that Chinese cities (see Table 7.8) could hold a total population of 567 million people to meet the estimated optimal density of 13,000 people/km^2 and a total built-up area size of 43,600 km^2. The actual total urban-area population is only 409 million people, so 12,100 km^2 of land could be saved under the optimal density. This means that reaching optimal urban population density could reduce total built-up area size by 27.84%, suggesting huge potential for land conservation. Enhancing the redevelopment of second-hand land and using land in an efficient and effective manner is a practical and feasible way to unleash the potential of land conservation.

Regionally, East China has the highest total urban-area population and built-up area size and the greatest potential for land conservation as a result of total regional area, urban area expansion and the ability to accept new migrants. In 2011, East China represented 47.80, 46.53 and 50.76% of these three measures respectively. Excluding municipalities directly under the Central Government, these measures adjust to 36.30, 39.79 and 48.81% respectively, which is much higher than China's other regions. Excluding data for municipalities directly under the Central Government,[8] East China has the lowest average urban population density of only 8559 people/km^2, with a regional land conservation potential of 5926 km^2. This figure,

[8]Municipalities directly under the Central Government were excluded from region- and urban population size-specific analysis unless otherwise stated. Given the uniqueness of these municipalities, they each have a large human population. Moreover, both Beijing and Shanghai are very populous and greatly affect data for East China and cities with populations over 2,000,000 people. As an example, East China and cities with populations over 2,000,000 people would each have a population density of 9637 and 11,173 people/km^2 respectively if municipalities directly under the Central Government were included. That compares with 8559 and 10,063 people/km^2 excluding such municipalities. For this reason, municipalities under the Central Government are excluded from region- and urban population size-specific analysis.

Table 7.8 Urban potential of land conservation under optimal urban population density, 2011

Indicator		Urban-area population (in million people)	Built-up area size (in km^2)	Urban population density (in 1000 people/km^2)	Urban potential of land conservation (km^2)	Percentage of potential of land conservation in built-up area size (%)
China		409.02	43,603	9.381	12,140	27.84
Region (excl. municipalities directly under the Central Government)	East	148.49	17,348	8.559	5926	34.16
	Northeast	46.72	5226	8.940	1632	31.23
	Central	84.10	8892	9.458	2423	27.25
	West	72.13	8162	8.837	2614	32.02
Urban population size (excl. municipalities directly under the Central Government)	>2.0 million	118.19	11,745	10.063	2653	22.59
	1.0–2.0 million	51.06	5960	8.567	2032	34.10
	0.5–1.0 million	69.35	7881	8.800	2546	32.31
	0.2–0.5 million	83.48	10,035	8.319	3613	36.01
	<0.2 million	29.37	4007	7.330	1748	43.62
Municipality directly under the Central Government	Shanghai	23.47	999	23.504	–	–
	Beijing	17.41	1231	14.137	–	–
	Tianjin	6.15	711	8.661	237	33.38
	Chongqing	10.43	1035	10.073	233	22.51

(continued)

Table 7.8 (continued)

Indicator		Urban-area population (in million people)	Built-up area size (in km^2)	Urban population density (in 1000 people/km^2)	Urban potential of land conserva-tion (km^2)	Percentage of potential of land conserva-tion in built-up area size (%)
Province and autonomous region	Guizhou	5.86	508	11.533	57	11.29
	Shanxi	10.12	957	10.579	178	18.62
	Qinghai	1.27	122	10.435	24	19.73
	Henan	21.59	2098	10.291	437	20.84
	Shaanxi	7.99	809	9.881	194	23.99
	Yunnan	7.94	804	9.879	193	24.01
	Hubei	17.82	1812	9.838	441	24.33
	Guangdong	47.35	4829	9.805	1187	24.58
	Liaoning	21.94	2277	9.640	588	25.85
	Hunan	13.38	1408	9.505	379	26.88
	Hainan	2.24	238	9.426	65	27.49
	Sichuan	16.69	1788	9.332	504	28.21
	Hebei	15.66	1685	9.294	480	28.51
	Fujian	10.33	1130	9.146	335	29.65
	Guangxi	9.08	1014	8.956	316	31.11
	Jilin	11.18	1271	8.800	411	32.31
	Jiangxi	8.56	1020	8.393	361	35.44
	Gansu	5.49	656	8.379	233	35.55
	Heilongjiang	13.65	1679	8.134	628	37.43
	Anhui	12.63	1598	7.902	627	39.21
	Inner Mongolia	8.41	1077	7.805	430	39.97
	Jiangsu	26.72	3494	7.647	1439	41.18
	Shandong	27.99	3751	7.463	1598	42.60
	Xinjiang	6.52	922	7.077	420	45.56
	Zhejiang	14.18	2221	6.386	1039	46.78
	Ningxia	2.37	371	6.376	189	50.95
	Tibet	0.49	90	5.446	52	58.11

along with its percentage of built-up area size of 34.16% is much higher than in any other region in China. West China has the second highest potential for land conservation and percentage of built-up area size at 2614 km^2 and 32.02% respectively. This relates to the weaker ability of Western Chinese cities to accept new migrants. Central and Northeast China had higher population densities but had the lowest percentage of land conservation potential and built-up area size, at 27.25 and 31.23% respectively.

With respect to urban population size,[9] small and medium-sized cities with populations less than 500,000 people have the lowest population density, at 7330 and 8319 people/km^2 respectively. Accordingly, these cities have the highest percentages of land conservation potential in built-up areas, at 36.01 and 43.62% respectively. This is due to the weaker ability of these cities to accept new migrants while their built-up areas are expanding. Ultra-large cities with populations over 2,000,000 people have the highest population density, or 10,063 people/km^2. Accordingly, they have the lowest percentage of land conservation potential in built-up areas of only 22.59%. This reflects the stronger ability of these cities to accept new migrants.

With respect to municipalities directly under the Central Government and provinces/autonomous regions, both Shanghai and Beijing have already exceeded optimal population size according to the optimal population density standards set in 2011. This is in part due to their strong ability as ultra-large cities to accept new migrants. The provinces of Guizhou, Shanxi, Qinghai, Henan and Shaanxi have the highest average urban population densities, and accordingly have the lowest percentages of land conservation potential in built-up area size. Apart from the autonomous regions of Inner Mongolia, Xinjiang, Ningxia and Tibet, which all have small populations, the provinces of Anhui, Jiangsu and Shandong have the lowest average urban population densities, and accordingly have the highest percentages of land conservation potential in built-up area size. In terms of the absolute amount of land conservation potential, cities in the provinces of Jiangsu, Shandong, Guangdong and Zhejiang have the highest potential for land conservation at 1439, 1598,1187 and 1039 km^2 respectively under the optimal urban population density standard. The total amount of conservable land in these four provinces represents about 43.35% of the total amount of such land in China.

7.6 Choices for Optimizing City Structures in China

According to analysis of the spatial expansion effect of Chinese cities, China's future urbanization needs to be conducted with resource efficiency in mind. The Chinese government should coordinate population and land urbanization under the constraints of optimal urban population density standards. Land idleness and inefficient use of land caused by poorly controlled urban sprawl in earlier stages of urbanization

[9]Total urban-area population herein equals urban-area population plus temporary population in urban areas.

is one of the main problems facing urbanization in China today. Both tight land supply in towns and cities and the great potential for land conservation indicates that resource-efficient urbanization is necessary and feasible. New urbanization with Chinese characteristics requires replacing the previous pattern of poorly controlled urban sprawl with resource-efficient urbanization driven by unleashing the potential of land conservation, increasing the ability of cities to accept new migrants and building more compact towns and cities.

1. Focus on unleashing the potential of land conservation and increasing the efficiency of land use in towns and cities: The current inefficient use of land in Chinese towns and cities is reflected in the building of infrastructure that exceeds its purpose and basic necessary standards, the inefficient use of land for industrial purposes, and the large amount of land sitting idle despite receiving approval for use. Accordingly, land conservation can be achieved by redeveloping second-hand land or using land in a more efficient and effective manner. This can be achieved through two parallel strategies. First, the government should establish a corresponding exit, penalty or redevelopment mechanism for idle or inefficiently used land. Such land can be correctly used by taking back land that violates agreed-upon conditions, encouraging land transfer, and reclaiming land already used for industrial purposes or using it for new industrial or non-industrial projects. Land can also be used more efficiently by rebuilding old factories, villages, towns and cities without damaging traditional cultures. Second, the traditional practice of attracting investment by providing low-cost land for industrial purposes must be changed. More efficient land use can be achieved by setting reasonable land prices and implementing higher requirements for acquiring land for industrial purposes.
2. Increase the ability of cities to accept new migrants and optimize population density in existing built-up areas: Apart from ultra-large cities in China's top three economic belts-the Pearl and Yangtze River Deltas and Beijing-Tianjin-Hebei region-where urban population density is high, most other Chinese cities have low population densities as a result of their weaker ability to accept new migrants. It is therefore necessary to increase the ability of Chinese cities—especially small and medium-sized cities- to accept new migrants so as to better coordinate development of towns and cities of differing sizes. Larger cities, especially very large and ultra-large cities, should enhance their service functions as regional centers, improve the industrial structure and increase the capacity of public facilities, thereby enhancing their overall carrying capacity. Likewise, towns and small/medium-sized cities should improve public services, enhance industrial support, guide people and industrial agglomeration, and increase the ability to accept rural migrants (The General Report Authoring Team 2013). Statistics show that small and medium-sized cities with populations less than 500,000 people have significantly lower population densities than other Chinese cities. This is a reflection of the weaker ability of these cities to accept new migrants. Increasing the ability of towns and small/medium-sized Cities to accept new migrants will reduce pressure on larger cities in terms of the agglomeration of people.

3. Build compact towns and cities to reduce crowdedness while increasing urban population density: Compact towns and cities are the most efficient city model in terms of land use. Compact cities are able to achieve economies of scale from agglomeration, while also reducing the crowdedness effect. The rightward shift of the density effect curve contributes to a higher optimal urban population density value when economies of agglomeration and crowdedness effect reach a balance. Compact towns and cities can be built by such means as high-density urban development and mixed-purpose land use and prioritizing the construction of a public urban transport system. High-density urban land development and use not only will significantly curb urban sprawl and effectively protect farmland and green spaces from development, but will also effectively reduce commuting distances and reliance on cars and thus reduce energy consumption and carbon emissions. Moreover, a compact pattern of development can accommodate more urban activities in a limited area, achieving economies of scale in investment in public services and facilities. Mixed-purpose urban land use refers to mixing residential, commercial, recreational and public service/facility purposes together to in order to limit commuting distances, thereby reducing transport demand and corresponding energy consumption and carbon emissions along with alleviating urban diseases. Building an urban transport system that prioritizes public transport will not only make this system operate more efficiently but will also help reduce reliance on cars and improve the urban environment by reducing car exhaust.

References

Alonso, W. (1970). The economics of urban size. *Papers of the Regional Science Association, 26,* 67–83.

Bettencourt, L. M. A. (2013). The origins of scaling in cities. *Science, 340,* 1438–1441.

Chen, L. W., Yang, K. Z., Shen, T. Y., & Wang, W. (2008). On how the density of economic agglomeration relates to differences in productivity. *China Economic Quarterly, 10,* 99–114.

Chris, C., & Jay, K. (2006). Controlling urban sprawl: Some experiences from Liverpool. *Cities, 23*(05), 353–362.

Ciccone, A. (2002). Agglomeration effects in Europe. *European Economic Review, 46*(2), 213–227.

Ciccone, A., & Hall, R. (1996). Productivity and the density of economic activity. *American Economic Review, 86*(1), 54–70.

Combes, P. P., Duranton, G., & Gobillon, L. (2008). Spatial wage disparities: Sorting matters! *Journal of Urban Economics, 63,* 723–742.

Cutler, D. M., Glaeser, E. L., & Jacob, L. V. (1999). The rise and decline of the American Ghetto. *Journal of Political Economy, 107*(3), 455–506.

Duranton, G., & Puga, D. (2004). Micro-foundations of urban agglomeration economies. In J. V. Henderson & J. F. Thisse (Eds.), *Handbook of regional and urban economics* (1st ed., Vol. 4, pp. 2063–2117). Amsterdam: North-Holland.

Duranton, G., & Puga, D. (2005). From sectorial to functional urban specialization. *Journal of urban economics, 57,* 343–370.

Ewing, R. H. (1994). Characteristics, causes and effects of sprawl: A literature review. *Environmental and Urban Issues, 21,* 1–15.

Fan, J. Y. (2006). On how industrial agglomeration relates to interregional differences in productivity. *Economic Research Journal, 11,* 72–81.
George, R. H. (2001). Just what is sprawl, anyway? *Carolina Planning, 26*(2), 11–26.
Glaeser, E. L., & Kahn, M. E. (2004). Sprawl and urban growth. In J. V. Henderson & J. F. Thisse (Eds.), *Handbook of regional and urban economics* (1st ed., Vol. 4, pp. 2481–2527). Amsterdam: North-Holland.
Henderson, J. V. (1974a). Optimum city size: The external diseconomy question. *Journal of Political Economy, 82*(2), 373–388.
Henderson, J. V. (1974b). The sizes and types of cities. *American Economic Review, 64*(4), 640–656.
Henderson, J. V. (1997). Externalities and industrial development. *Journal of Urban Economics, 42*(3), 449–470.
Henderson, J. V., Ari, K., & Matt, T. (1995). Industrial development in cities. *Journal of Political Economy, 103*(5), 1067–1090.
Kahn, M. E. (2000). The environmental impact of suburbanization. *Journal of Policy Analysis & Management, 19*(4), 569–586.
Krugman, P. R. (1991). Increasing returns and economic geography. *Journal of Political Economy, 99*(3), 483–499.
Li, Q., Liu, A. G., & Zhu, H. C. (2005). A review of western studies on urban sprawl. *Foreign Economics & Management, 10,* 51–58.
Liu, X. Y. (2009). On how economies of agglomeration relate to productivity. *The Journal of Quantitative & Technical Economics, 7,* 109–119.
Marshall, A. (1890). *Principles of economics*. London: Macmillan.
Mills, E. S. (1975). An aggregative model of resource allocation in a metropolitan area. *American Economic Review, 57*(5), 197–210.
Mills, E. S., & Ferranti, D. M. (1971). Market choices and optimum city size. *American Economic Review, 61*(2), 340–345.
Montgomery, M. R. (1988). How large is too large? *Economic Development and Cultural Change, 36*(4), 691–720.
Newman, P., & Jeffrey, K. (1998). *Sustainability and cities: Overcoming automobile dependence*. Island Press Books.
Putnam, R. D. (1999). *Bowling alone: The collapse and revival of American Community* (p. 2000). New York: Simon & Schuster.
Segal, D. (1976). Are there returns to scale in city size? *Review of Economics and Statistics, 58*(3), 339–350.
Sveikauskas, L. A. (1975). The productivity of cities. *Quarterly Journal of Economics, 89*(3), 393–413.
The General Report Authoring Team. (2013). The overall strategy for promoting the hukou urbanization of rural migrants. In J. H. Pan & H. K. Wei (Eds.), *Annual report on urban development of China No. 6* (pp. 1–43). Beijing: Social Sciences Academic Press. (China).
The Sustainable Development Strategy Study Group of the Chinese Academy of Sciences. (2009). *China sustainable development strategy report 2009*. Beijing: Science Press.
Tu, G. X., Wang, Z., & Zhang, Z. L. (2013). A period of tight land supply in China: Increasing land supply in four ways. *China Comment, 14,* 6–17.
Wang, J. T., Zhao, L., Feng, S., & Zhao, Y. J. (2014). On how urban sprawl manifests itself and affects the environment. *Urban Problems, 5,* 22–27.
Wei, H. K. (2014). *A new road to urbanization with Chinese characteristics*. Beijing: Social Sciences Academic Press (China).
Wei, H. K., & Su, H. J. (2013). On the hukou urbanization of rural migrants in China. *Chinese Journal of Population Science, 5,* 21–29.
Wei, H. K., & Wu, X. X. (2009). China's anti-poverty policy: Comment and outlook. *The Journal of Shanghai Administration Institute, 2,* 56–68.
World Bank. (1993). *China: Strategies for reducing poverty in the 1990s*. Beijing: China Financial & Economic Publishing House.

Chapter 8
Recommendations for Comprehensive Improvements in Urban Development and Management

China has made great achievements in urban development. Some of China's larger cities are already world-class in terms of infrastructure. Nonetheless, China's pattern of urban development has overlooked management and quality, causing a series of problems including the loss of urban features, poor management, the emergence of shanty towns and the unreasonably rapid expansion of very large and ultra-large cities. At the Central Urbanization Work Conference in 2013, the Chinese government announced that one of the main tasks for future urbanization would be improving the quality of urban development. Taking reality and policy context into account, future efforts for new urbanization should focus on addressing existing problems facing urban development and management in China and taking measures to improve both aspects, thereby facilitating sustainable urban development and achieving high-quality, healthy urbanization.

8.1 Improving the Quality of Urban Development

Chinese cities are quickly approaching or on par with the world's best cities in terms of physical infrastructure. However, they face three key challenges (see Table 8.1). First, Chinese cities are so devoid of local features to the extent that many of them look alike. Most Chinese cities have lost or are lacking cultural, ethnic and local characteristics. This is due to a lack of awareness of cultural heritage, low-quality urban plans and designs, and a habit-copying urban planning and development. Chinese cities are gradually becoming similar regardless of their location and size. The second challenge is the existence of low quality buildings with short life spans, caused by ineffective building quality management and a high rate of housing demolition arising from frequent changes to urban plans. One scholar has noted that despite China's 5000-year-old civilization, only a small number of buildings are more than 50 years old (Yang 2013). Residential buildings in Chinese cities tend to last for only 25–30 years, much shorter than the average of 132 years in the United King-

H. Wei, *Urbanization in China*, Research Series on the Chinese Dream and China's Development Path, https://doi.org/10.1007/978-981-13-1408-7_8

Table 8.1 Primary problems with urban development in China

Problem	Cause	Symptom	Countermeasure
No local features	Lack of cultural inheritance	Loss of local features due to damage to historic and cultural sites	Enhance cultural inheritance and site protection
	Low-quality planning and design	Very similar building designs; excessive ancient- and foreign-style buildings	Improve plans/designs; curb the practice of copying
	Competition for grandeur among local governments	Extravagant government buildings, skyscrapers sand squares	Stop the focus on competition; emphasize local features
Low-quality buildings with short life spans	Low building quality	Jerry-building	Quality management and supervision throughout the process
		High percentage of conventional building materials; high resource and energy consumption	Develop green buildings; industrialization of residential buildings
	Demolition due to plan changes	Frequent changes due to changes in local leadership	Make plans more scientifically-sound and sustainable
Prioritizing ground infrastructure over underground infrastructure	Insufficient attention; false idea	Frequent urban flooding due to an ineffective drainage system	Emphasize and enhance underground piping

dom and 74 years in the United States. Third, ground-level infrastructure receives much more attention in Chinese cities than underground infrastructure. This is a common problem with urban development across China. While high buildings and ground infrastructure are constructed at a large scale, underlying infrastructure like the drainage system is neglected. The severity of this problem is evident in incidents such as the flood that hit Beijing on July 21, 2012[1] after torrential rain. Urban flooding has occurred in over 300 of China's 500 cities, according to incomplete statistics (Sun et al. 2012).

The current problems with urban development in China are primarily problems with urban planning and building construction. On the urban planning and design side, some local governments do not value historic sites and cultural resources, and have systematically damaged buildings of historical and cultural significance in order

[1]The heaviest rain and flooding in China in the past 61 years hit Beijing and its surrounding areas on July 21–22, 2012 at around 8:00 am, causing a death toll of 79 people, 10,660 collapsed houses, 1,602,000 victims and a financial loss of 11.64 billion yuan.

to rapidly profit from land. In so doing, they have reduced the cultural value of such buildings by making them less compatible with the surrounding area during urban development, or by developing them in a predatory manner in the name of promoting tourism (Qiu 2012). In addition, city quality has been adversely affected by the practice of copying urban plans and designs from other cities, using the same set of drawings in different projects to build identical copies of ancient- or foreign-style buildings. In previous years, extravagant government buildings, skyscrapers, squares and other structures flourished in China as local governments sought to compete with one another in terms of grandeur. While a lot of money has been spent on implementing these highly visible, white elephant projects, local governments have failed to pay sufficient attention to underground infrastructure. Unreasonably low drainage system and flood prevention standards have led to poor quality underground infrastructure. Even in nationally important central cities such as Guangzhou, 83% of all drainage pipes in central urban areas can only sustain one-year floods; only 9% of such pipes can handle two-year floods. In addition, frequent changes in urban plans caused by changes in municipal leadership have led to a lack of sustainable urban development. This is exemplified by the frequent demolition of new buildings, which in turn results in a shorter average life span of all buildings. The 131-meter-high Millennium Tower in Hainan Province was less than 10 years old when it was demolished due to changes in the local urban plan, for example. Longsheng Tower, a building in Chongqing City, was demolished just six months after it was delivered, once again due to changes in the local urban plan. Furthermore, jerry-building and even building collapse occurs from time to time as a result of ineffective supervision.

For future urbanization, a full range of measures is needed to combat the problems with urban development and address their root causes in order to effectively improve urban development and enhance local features and character, with a focus on city quality.

1. **Change the mindset in terms of the value of urban cultures and green development**. The lack of cultural inheritance and mechanical development of natural elements are important causes for the similar appearances of Chinese cities. Future urban development should value cultural heritage while adding modern elements. The Chinese government mindset toward urbanization must change to include the protection of buildings of historical and cultural significance into the government performance assessment system. It is necessary to improve the existing governing laws and regulations—such the *Urban and Rural Planning Law*, the *Cultural Relics Protection Laws* and the *Regulations on Protection of Villages, Towns and Cities of Historical and Cultural Significance*—to create a complete legal system that protects buildings, towns and cities of historical and cultural significance. The demolition of any historic building must be denied as long as there is only one vote against it. The government should promote people-oriented city development and more rapidly enact national plans and standards for building such cities. Measures should be taken to promote the building of people-oriented cities as demonstrations. In addition, cities should be built as an integral part of nature. The building of green cities should be acceler-

ated, along with demonstration eco-friendly cities/districts. Urban development should be based upon preserving viewscapes of natural elements such as rivers, lakes and mountains. In addition to respecting and protecting nature, natural elements should be incorporated into city design wherever possible.

2. **Improve urban plans/designs and plan management**. Problems with urban planning are an important reason for the lack of quality and short lifespans of urban buildings. It is therefore necessary to improve urban plans/designs and plan management. Concepts such as a people-centric approach, respect for nature, cultural inheritance and green development should be embodied throughout urban planning and design. It is critical to help urban planners and designers become more skillful and innovative to minimize the occurrence of design copying. It is also necessary to change the mindset regarding urban planning and design so as to improve the planning, design and building of underground infrastructure such as drainage pipes. With respect to plan management, it is necessary to avoid arbitrary changes and shortsightedness of urban plans. In other words, urban planning should be more consistent and scientifically sound, and urban plans should not change simply as a result of changes in local leadership. Efforts should be made to improve procedures ranging from early research, planning, coordination, review by experts, public engagement and approvals to implementation management and assessment/modification. It is also necessary to make urban planning more transparent by enhancing public engagement and scrutiny.
3. **Improve building quality**. Ineffective management and low-quality building materials are the of building for low building quality in China. To improve building quality, several steps are needed. The first step is to enhance building quality regulations. Measures in this regard include strengthening quality control throughout the construction process from design, to supervision of building materials, to construction, to decoration/remodeling. Second, urban development should be conducted in successive steps starting with surveys, then proceeding to design, then to construction. The simultaneous implementation of these steps should be avoided. The third step is enhancing the management of qualifications for players in the construction market. This can be accomplished by promoting relevant quality certification systems and increasing professional skills training for construction workers. Stronger measures must be taken to combat illegal behavior related to invitations to bid, bidding, materials procurement, subcontracting and implementation of construction projects. It is necessary to improve the building registration record and inquiry and management systems as well as strengthen accountability and penalties for building quality. Implementing a lifelong accountability system is especially important. Along with improving building quality, greater effort should be made to develop new building materials, promote green buildings and industrialize residential buildings. New building materials should be developed that take less energy to produce and reduce energy consumption of buildings. Green measures that promote energy efficiency of buildings should also be accelerated, including the development and application of new walls, waterproofing, sealing, and applying thermal insulation. Measures

should be taken to promote green buildings while improving green technologies and management methods. Prefabricated residential buildings should also be encouraged in the industrialization of such buildings.

8.2 Modernizing Urban Management

Chinese cities are on a par with or even exceed foreign cities in terms of physical infrastructure, yet they lag far behind in management. Urban management in China does not receive sufficient attention in the urbanization process, and cities currently have not made significant progress in comprehensive management, managerial structure, details and emergency management and management methods. Urban management in China today is riddled with problems, especially in areas affecting daily life, such as cyclist safety, availability of taxis and parking spaces, mobility challenges for the disabled, problems with bus stations, city administration and unauthorized charging for parking. In terms of emergency management, Chinese cities are increasingly vulnerable in the midst of rapid expansion. In an emergency situation, urban operations would be affected by problems in a range of areas such as water/power/gas supply, transport, communications and logistics. Emergency incidents such as the outbreak of the severe acute respiratory syndrome (SARS) between 2002 and 2003, disasters caused by freezing rain and snow in southern China in early 2008, and the torrential rain that hit Beijing in 2012 have provided a wake-up call to Chinese cities. In terms of comprehensive management, the increasing complexity and size of the urban system is making it more difficult for specialized functional management to meet the ever increasing requirements. Managing the urban transport system, for example, involves municipal infrastructure, landscaping, sanitation, underground piping, power supply to traffic lights, a water service system that connects with the roadside drainage system, and so forth. In terms of managerial structure, as city sizes continue to increase there is a growing need for a managerial structure that is based on public engagement and an urban management structure based on digital grids. Given the current problems with urban management in China, efforts to modernize urban management for future urbanization should focus on changing the philosophy of urban management, managerial structure, priority and management methods (see Table 8.2).

1. **Shift from specialized management to comprehensive management**. Early urban development has led to the specialization of urban management characterized by subsystems specific to municipal infrastructures, public utilities, transport management, waste management, city appearance and landscape management and environmental management. As the complexity of the urban system grows, however, the connections between urban management subsystems have become too complex to operate individually. In other words, the task-specific, simplified pattern of specialized management no longer effectively addresses issues in urban

Table 8.2 Traditional versus modern urban management

	Traditional	Modern
Philosophy	Emphasis on specialized management	Emphasis on comprehensive, systematic management
Structure	One-way management	Participatory management
Priority	Managing urban operations	Paying equal attention to operations, details and emergency management
Method	Based on temporary campaigns	Based on digital grids

management.[2] A series of problems such as fragmentation, isolation, functional overlap, managerial inefficiency and ineffectiveness and lack of coordination have become common in contemporary urban management across China (Song and Tang 2007). The development of the urban system into a complex, overarching system (Zhou 2002; Zhai et al. 2011; Song and Tang 2007) depends not only on the proper development of subsystems but more importantly on coordination among subsystems for overall management. To this end, it is necessary to enhance research on comprehensive urban management and set up a corresponding comprehensive management department and coordination mechanism.

2. **Shift from one-way management to participatory management**. The traditional pattern of one-way urban management is increasingly unsuited to China's current urban society, which is diversifying with an increasingly layered social structure and an emerging variety of interest groups (Liu 2009). In 1986, the UNDP, UN-HABITAT and the World Bank co-launched the Urban Management Program (UMP) in the field of participatory urban governance. This program was designed to focus on urban issues affecting rapidly urbanizing developing countries, such as urban environment, poverty and infrastructure supply. Participatory management is a necessary choice for making urban management more effective and efficient, scientifically sound, democratic and people-oriented. Systematic participatory management is supposed to cover urban planning, development, operation and management. Its priorities include enhancing the development of non-governmental organizations (NGOs), improving community services and building sound communication mechanisms. To enable a larger society to better manage itself, effective measures should be taken to develop NGOs, especially non-profit NGOs that assist in the provision of services, including government services. It is also necessary to improve community functions relevant to public services and accelerate the introduction of these services into communities.

[2]Aside from the aforementioned problems with the transport system, urban managers who wish to achieve absolute control over urban water services must take into account that the urban water service system actually interacts with the urban infrastructure system, river/lake ecosystems, public utility system and even underground pipes for power supply. Communications cables and heating pipes area also laid in parallel with water supply/drainage pipes as well as with the roads above them.

Specifically, measures should be taken to integrate managerial functions and service resources into the fields of population management, employment, social security, civil affairs, health and birth control, cultural services, comprehensive governance, social stability maintenance and complaints handling. It is advisable to accelerate the deployment of community-relevant IT systems so as to build an integrated service and management platform for each community. In addition, it is necessary to create more convenient and efficient ways of communication between urban managers and citizens, such as an exchange mechanism, to enable a positive atmosphere under which citizens are willing to actively participate in urban management.

3. **Pay equal attention to normal operations management, details management and emergency management**. Besides maintaining normal urban operations, it is necessary to enhance details management and emergency management. With respect to details management, municipal governments should prioritize and refine the management of issues closely connected with people's daily lives. They may rely on participatory management, in particular, to pool the brain trust of all stakeholders in order to address these issues. They may also rely on grid-based management to achieve objectives regarding refined management. With respect to emergency management, it is necessary to build a sound urban emergency management system to guard against urban vulnerabilities. An emergency management system is an important component of modern urban management, the goal of which is to warn against, mitigate and even eradicate threats to urban public safety/security, such as natural disasters, accidents and public health and security incidents. To build such a system, municipal governments should enhance the building of facilities and capacities relevant to firefighting, flood prevention, drainage and earthquake relief. They should create more stringent disaster prevention standards for buildings, and reasonably locate and build emergency infrastructure such as emergency shelters, medical care, logistics and transport systems. Municipal governments should also undertake a variety of measures including: improving disaster monitoring and warning systems; enhancing disaster prevention and mitigation capacity; improving the administrative accountability system; improving contingency plans for sudden public incidents and improving emergency support systems; enhancing disaster analysis and information disclosure; educating citizens on how to prevent risks and protect themselves and others; establishing a catastrophe insurance system to allow the private sector to play an important role in emergency management.
4. **Promote grid-based management**. Grid-based urban management is the generic term for a whole set of urban management concepts, methods, organizations and processes that are built upon digital technology and characterized by grid cell management (Jiang and Ren 2007). It focuses on achieving digitized, closed-loop, refined, dynamic and real-time urban management. In 2004, Dongcheng District, Beijing City was the first region in China to conduct grid-based urban management. It has since produced great result sand has been gradually adopted in other districts and cities. Many larger Chinese cities have

already implemented grid-based management, and the experience shows that this type of management can make urban management much more efficient and less expensive, and thus is conducive to establishing a long-term, effective urban management mechanism that incorporates participatory management and regulates the behavior of urban managers (Jiang and Ren 2007; Chi et al. 2008). With that said, some Chinese cities have experienced problems with grid-based management, such as an inefficient mechanism, an incompetent team, insufficient investment, outdated hardware/software and insufficient promotion in implementation. In order to implement grid-based management across the board, it is necessary to improve management methods, tools, organizational structures and management processes. in the near term, effort should be taken to improve technologies such as remote sensing, geographic information system (GIS), global positioning system(GPS), distributed databases and computing, grids and grid computing, as well as components and component libraries. It is important to use grid cell and urban component management methods and enhance the use of information collection tools. In terms of organizational structure, it is necessary to establish separate supervision and management organizations. With respect to management processes, it is necessary to establish and improve the information collection, case record generation, task assignment, task handling, feedback, verification/closing and overall assessment systems.

8.3 Addressing the Dual Spatial Structure of Cities

As mentioned in Chap. 7, one of the main achievements of China's urbanization is the general lack of slums that are common in many other developing countries. With that said, it should be noted that there is growing residential space differentiation and an emerging dual spatial structure in Chinese cities arising from fast urbanization. This dual spatial structure—a forest of skyscrapers in certain areas and a sea of hovels in other areas—has become a tough challenge for urban development and management in some Chinese cities, especially old industrial centers and larger cities. The dual spatial structure of Chinese cities is mainly manifested by shanty towns, "villages in the city" and poorly maintained old urban areas that contrast starkly with modern, new, central urban areas. Such a dual structure is unique to Chinese cities. These urban areas, especially shanty towns, will likely devolve into slums without comprehensive improvement measures. The path to new urbanization with Chinese characteristics requires accelerating the rebuilding of shanty towns, the comprehensive improvement of villages in the city and the functional improvement of old urban areas. These measures will radically change the dual spatial structure of Chinese cities

8.3.1 Promoting the Rebuilding of Shanty Towns

Houses in shanty towns provide the worst residential conditions in China. These houses came into being by three main means: they were built by employers for factory workers or poor people before the founding of the People's Republic of China in 1949; they were built by employers in or around factories, mines and forestation/farming zones after October 1949; or they are makeshift dwellings built by residents with or without government approval (Zhang 2010). Houses in these shanty towns are crude and aging structures with low resistance to disasters, poor functionality, low-quality living environments, lack of spaces for public events and, and lack of supportive facilities. Moreover, households living in shanty towns are generally low-income or lower-middle-income households that cannot afford housing.[3] Shanty towns in China are similar to slums in other countries in terms of residential conditions and resident income. The presence of shanty towns has produced significant adverse effects on the city's appearance while also isolating residents in shanty towns from those in surrounding communities. This dual spatial structure reflects the growing urban social differentiation in China.

Rebuilding various shanty towns is important not only in improving a city's image but also in addressing the dual structure. China began large-scale rebuilding of shanty towns in 2005. shanty town rebuilding has become an important program for municipal governments to improve public welfare and win public trust, particularly since the Central Government resumed the affordable housing program in 2008. Nearly a decade of large-scale shanty town rebuilding has produced local experience and patterns, such as the Fuxin pattern driven by the transformation of a resources-rich city, the Changsha pattern led by better housing assurance and the Harbin pattern led by a shift from commercial development toward subsidization by the government (Zhang 2010). Despite such experience, there remain perennial problems due to the scale and difficulty of rebuilding. Between 2008 and 2012, 12.60 million houses in shanty towns across China were rebuilt, but over 12 million low-income or lower-middle-income urban households still live in shanty towns, according to the *Opinion of the State Council on Accelerating Shanty Town Rebuilding* (GF [2013] No. 25). This policy document sets a goal of rebuilding 10 million more houses in shanty towns between 2013 and 2017. Unfortunately, however, there are significant difficulties in rebuilding shanty towns, including difficulty related to housing demolition and resettlement as a result of poor primary and auxiliary infrastructure.

Future shanty town rebuilding should prioritize shanty towns in cities, state-owned factories and mines, and afforestation/farming zones. Under the people-oriented principle, shanty town rebuilding should be transparent in terms of demolition and reset-

[3]By the end of 2008, there were 11.48 million households living in shanty towns across China, including 7.44 million households in cities (64.80%), 2.38 million households in state-owned factories and mines (20.73%) and 1.66 million households in forestation/farming zones (14.46%); of these, 6.81 million households were low-income and could not afford housing and 4.67 million households were lower-middle-income and could not afford housing, according to a survey. There remain about 12 million households living in shanty towns due to adjustments to China's shanty towns standards. The percentages of various shanty towns remain stable.

tlement, should use quality materials in construction, and promote the standard of living for the affected residents. To ensure a reasonable standard of living for the affected people, it is necessary to build affordable housing in shanty towns, create a fair and reasonable dwelling replacement mechanism, and improve late-stage support services such as community services, property management and maintenance. With respect to transparency in demolition and resettlement, it is necessary to improve the pricing and compensation mechanisms for land expropriation, demolition and resettlement in shanty towns. The government should increase subsidies for areas with lower development value so as to reduce concerns of affected households. Forced demolition and resettlement must be prohibited. With respect to quality assurance in construction, shanty town rebuilding should be properly planned to construct new, high-quality settlements with access to a full range of infrastructure including water/power supply, drainage, transport, MSW treatment and communications, as well as access to commercial facilities. In addition, competent industry authorities should enhance supervision and guidance of shanty town rebuilding in state-owned factories and mines.

8.3.2 Enhancing Governance of "Villages in the City"

Given the costs of housing demolition and villager resettlement, municipal governments or developers have preferred acquiring low cost farmland for urban development, avoiding larger villages. Consequently, a large number of villages have remained amidst urban expansion, and are known as "villages in the city". The issue of "villages in the city" is the most severe in larger cities such as Beijing, Tianjin, Chongqing, Shanghai, Wuhan, Guangzhou and Shenzhen. There are three main problems with the status of villages in the city. First, villagers usually build houses without permission due to lack of authority over land management and urban planning. Many such villages in the city have building densities of over 70% and even as high as 90%. Consequently, "shake-hand buildings", "face-to-face buildings" and "a thin strip of skies" (Feng 2014) have become common phenomena in Chinese cities. Second, such low-quality buildings provide non-locals with dwellings available at low rents, resulting in a high percentage of non-locals residing in villages in the city. This creates challenges related to high mobility, a complex resident mix and security. Third, villages in the city generally cannot benefit from municipal government infrastructure development programs since they are not covered by urban plans. Moreover, villagers or village committees are unable to build infrastructure on a large scale due to problems with land ownership, planning power and funding. Consequently, infrastructure related to transport, water/power supply, sewage drainage and MSW treatment in villages in the city are far from sufficient.

The current rebuilding of villages in the city is carried out under three primary patterns, government-led rebuilding, village-led rebuilding, and cooperation between developers and villages (Tao 2012). Under the first pattern, local governments play a leading role in making policies specific to rebuilding, compensating affected residents for housing demolition and resettlement, and drafting and implementing villager resettlement plans. Under the second pattern, the village conducts rebuilding on a joint stock basis and raises funds to cover demolition, resettlement and building construction. Under the third pattern, villages identify and cooperate with developers to rebuild.

The type of pattern used for rebuilding depends on the rights to collectively-owned assets and curtilage in villages in the city. A village in the city without collectively-owned assets in urban built-up areas can basically be rebuilt in the same way as shanty towns are rebuilt. Villages in the city with both collectively-owned assets and curtilage face greater difficulties in rebuilding. Villagers in well-located villages in the city rely on renting out their homes for a large portion of their income, and some of villages also share profits from collectively-owned assets and village-run enterprises. Accordingly, it is difficult to rebuild these villages under the government-led pattern. Since land ownership is the root cause of problems related to rebuilding these villages in the city, it is important to recognize the rights of collectively-owned economic organizations and rights of farmers with respect to long-term development and incomes on the land used for rebuilding (Li 2013). When the costs of forced demolition/resettlement and compensation are unfeasible, villages should be encouraged to conduct rebuilding for themselves or cooperate with developers. It is inappropriate to expropriate land within these villages and turn it into state-owned urban land. Instead, affected farmers may retain ownership of the original collectively-owned corporate assets. For residents affected by land expropriation, they may also become shareholders on a pro-rata basis or rent out their shares of the land as an ensured source of income.

8.3.3 Making Functional Improvements to Old Urban Areas

Compared to the rebuilding of shanty towns and villages in the city, which is an issue unique to China, the renewal of old urban areas can be regarded as a universal issue in the process of urban development. There is a long history of renewal of old urban areas outside China. The urbanization rates of European and American countries have been high and stable since the mid-20th century. The subsequent pattern of urban development focuses on the renewal and improvement of old urban areas (Zhang 2009). Accordingly, these countries have a great deal of experience in the renewal of old urban areas. In 1958, the first international conference on the renewal of old urban areas was held in The Hague, the Netherlands. The conference defined such renewal as planned renewal in old urban areas focusing on redevelopment, restoration and protection in accordance with urban development requirements (Xiang 2005). There are two aspects of the experience of developed countries that deserve closer study. First is the focus of urban renewal on improving municipal infrastructure and the living environment. Second is the combination of the renewal of old urban areas with the protection of their historic and cultural features. Important cultural and historical sites and objects, particularly ancient buildings, should be well protected and their original design maintained (Xiang 2005).

As Chinese cities develop, four types of old urban areas have emerged: central urban areas, historic and cultural sites, multi-purpose areas and industrial areas. Based on international experience in the renewal of old urban areas, renewal patterns should be adapted to reflect the type of old urban area. Central urban areas have advantages such as established infrastructure, centralized support facilities and good location, but have disadvantages such as high building density, a small amount of public green space, a poor living environment and insufficient parking spaces. These areas should be renewed to address their strengths and weaknesses. The focus should be on transforming these areas into central business districts (CBDs) with historic buildings and landscape features preserved. Historic and cultural sites, such as the *siheyuan* in Beijing (also referred to as Chinese quadrangle), the Bell and Drum Towers in Xi'an, and the Confucius Temple in Nanjing were heavily damaged in early urban development. The restoration of historic and cultural sites should focus on preservation and cultural inheritance. Multiple-purpose areas are a mixture of residential, commercial, industrial and municipal facilities usually located in the middle circle of a city. A product of shortsighted early urban plans, these areas have narrow roads and dense buildings. Accordingly, it is more difficult to conduct demolition in these areas. Given the high residential building density and poor location of these areas, combined with difficulty in raising funds, redevelopments of these areas depends on bundling multi-purpose areas and well-located pieces of land together for redevelopment. For industrial areas, it is generally necessary to adjust the industrial distribution in a city to reflect urban industrial structure improvements caused by urban expansion and functional changes. The redevelopment of industrial areas is usually based on industrial transformation and involves changes in employment and population structures. In other words, it is essentially regional socioeconomic transformation. It is therefore necessary to develop a scientifically-sound socioeconomic transformation strategy to renew industrial areas while at the same time protecting and utilizing industrial heritage sites.

8.4 Addressing Urban Sprawl

Larger Chinese cities, especially very large and ultra large cities, have been expanding rapidly in terms of population and land size in recent years. This rapid expansion is due to resource allocation bias in favor of administrative centers and larger cities, combined with the polarization effect arising from mutual reinforcement between economies of agglomeration and market forces. This is highlighted by urban sprawl and the continuous increase in built-up areas. In 2012, there were 13 Chinese cities with an urban-area population of over four million people, excluding the temporary population. Of these, Shanghai, Beijing, Chongqing, Shenzhen and Hangzhou had populations of over ten million people. Except for Dongguan, where the temporary population accounted for nearly 70% of its urban-area population, these 12 cities saw a 27.4% average increase in urban-area population size and 55.5% increase in built-up area size between 2006 and 2012. This was 2.06 and 1.57 times higher than the average growth rates of cities nationwide, respectively. In this period, these 12 cities saw their combined percentage in the total urban-area population of all Chinese cities rise from 23.7 to 26.6%, up 2.9 percentage points. 49% of all new urban residents across China moved to these 12 cities (see Table 8.3).

The rapid expansion of larger cities has caused problems including crowdedness, traffic congestion, small per-capita housing area, resource shortages, environmental pollution and worsening poverty. The population density is 24,200 people/km^2 in Shanghai's central urban areas and 23,400 people/km^2 in Beijing, both of which are higher than the population densities in the central urban areas of Paris, Tokyo and New York, according to estimates (Wang et al. 2012). Traffic congestion is a common problem in all larger cities in China, especially very large and ultra-large ones, to the point where traffic in central urban areas is practically paralyzed in rush hour. Traffic congestion has significantly increased commuting time in larger cities. A survey shows that the average commuting time to and from work already exceeds one hour in cities such as Beijing (1.32 h), Shanghai (1.17 h), Tianjin (1.15 h), Shenyang, Xi'an and Chengdu (Zhaopin.com and the Peking University Center of Social Survey Research 2012). Other common problems in these larger cities include high and rapidly increasing housing prices, insufficient ecological spaces, severe environmental pollution and intensified social conflicts. Globally, it is widely believed that a reasonable ratio of housing price to income is between 4 and 6. However, in 12 of China's Tier 1 cities—Shanghai, Beijing, Chongqing, Tianjin, Guangzhou, Shenzhen, Chengdu, Hangzhou, Suzhou, Nanjing, Wuhan and Shenyang-the average ratio of housing price to income was 11.04 in 2010, increasing from 8.65 in 2005 (Ding 2013). Between 2007 and 2012, thirty-five medium-sized and large cities across China saw the average housing price rise by 109%, which was much higher than the growth rate of urban residents' average disposable income in the same period.

Table 8.3 Urban-area population growth of 12 ultra-large Chinese cities

City	2006 (million)	2012 (million)	Population Growth (%)	City	2006 (million)	2012 (million)	Population growth (%)
Shanghai	18.1508	23.8043	31.15	Shenyang	4.5761	5.7136	24.86
Beijing	13.3300	17.8370	33.81	Nanjing	4.3132	5.6727	31.52
Guangzhou	9.8554	10.1500	2.99	Zhengzhou	2.6120	5.9166	126.52
Shenzhen	8.4643	10.5474	24.61	Chengdu	3.9024	4.5831	17.44
Chongqing	8.324	11.1830	34.32	Harbin	4.1525	4.3061	3.70
Wuhan	4.9300	6.2752	27.29	Subtotal	88.2866	112.4831	27.41
Tianjin	5.6745	6.4941	14.44	All Cities	372.7280	422.2680	13.29

Note Urban-area population includes temporary population. Dongguan City is not included due to its large temporary population and, hence, an unstable urban-area population
Source China Urban Construction Statistical Yearbook (2006 and 2012)

Comprehensive measures must be taken to address the increasingly visible problems caused by the expansion of larger cities. First, the Chinese government should keep tight control over the population sizes of very large and ultra-large cities. The root of the aforementioned urban problems is excessively large city size or rapid expansion that exceeds the overall carrying capacity of the city. For very large cities with populations of at least one million people, it is necessary to implement population size control policies depending on the specific situation. Today, very large Chinese cities with populations between one and two million people still have relatively high development potential; cities with populations between two and four million people face increasing limitations in terms of local resource and environmental carrying capacity; and cities with populations over four million people are already overcapacity. Accordingly, population control measures should target cities with populations over four million people. In addition to adopting a points-based system to restrict household residence registration, the key to population size control is to develop reasonable industry entry standards and optimize the industrial structure according to the functional positioning and developmental orientation of these cities. This will help ultra-large cities develop high-end industries and services and reduce population pressure through industrial transformation and optimization.

Second, the Chinese government should implement a functional and industrial decentralization strategy. Globally, a common practice in addressing problems caused by large city sprawl is to build satellite towns, limit the building of new facilities in central urban areas and encourage companies to move out. It is therefore necessary to implement a functional decentralization strategy according to the overall carrying capacity and functional positioning of Chinese cities. Competent authorities should provide reasonable guidance to larger Chinese cities in shifting people, industries and facilities from central urban areas into the suburbs and surrounding regions in accordance with the functional decentralization strategy. This will promote the structural optimization of space in these cities, regional integration and sustainable development. Functional decentralization is the key to controlling the population size of larger Chinese cities.

Third, the Chinese government should increase the overall carrying capacity of very large and ultra-large cities. To address the insufficient capacity of public infrastructure, the government should increase the supply of public service infrastructure in larger cities, especially for ordinary people including particularly disadvantaged groups. Priority should be given to the development of public transport, particularly to the building of high capacity, rapid rail transport, bus rapid transit (BRT) and inter-city rapid rail transport systems. Efforts should be made to build urban public transport networks that are convenient, safe, economical, comfortable and green. The use of public and green transport should be encouraged, and staggered working hours should be adopted. Competent authorities should endeavor to build public transport-based cities. For larger cities urban-area populations of at least one million people, rail transport should be planned to alleviate traffic congestion. Globally, rail transport accounts for 65% of the passenger traffic in Paris and London, 61% in New York and 86% in Tokyo (Yuan 2005). In addition, for larger cities with severe water shortages, measures such as inter-regional water diversion and seawater desalination that meet technical and economic evaluations as well as environmental impact assessments (EIA) should be explored to increase carrying capacity.

Fourth, the government should enhance the ability to deal with urban diseases. In order to address major pollution problems related to air and water quality and the MSW, three steps are needed: develop and implement specific action plans; establish or improve statistical, monitoring, warning and assessment systems; and improve regional joint prevention and control mechanisms for reducing the total amounts of primary pollutants. These steps will enhance comprehensive action to combat environmental pollution in larger cities. Competent authorities should also manage vehicle supply and demand. Measures such as license plate auctions, limited use and one-way streets, increased prices for parking and road use, exclusive public transport lines and improved traffic management should be undertaken. Increasing the cost of vehicle ownership and use, controlling the amount of vehicle travel, optimizing traffic flows and distribution, and enhancing the ability to guide traffic will alleviate or even eliminate traffic congestion in larger cities.

8.5 Building New Types of Cities Across China

Building new types of cities is an important approach to improving urban development and management. The *National New Urbanization Plan (2014–2020)* calls for the building of green cities, cultural cities, and smart cities. Existing demonstration urban development projects in China include National Garden Cities, forest cities, next-generation Internet cities, pilot smart cities, low carbon cities, low-carbon eco-cities, ecological civilization cities, national pilot demonstration zones for ecological civilization, cities with circular economies, and cities powered by new energy sources (see Table 8.4). Existing demonstration cities primarily emphasize intelligence and eco-friendliness, but do not yet emphasize culture. In order to promote new urbanization, it is necessary to fuel the development of cultural cities by consolidating various green and smart city building programs. Note that these three types of cities are not mutually exclusive—eco-friendliness, intelligence and culture should be equally valued in the development of a city. The Beijing Municipal Government, for example, announced the goal of building a "green, cultural and high-tech Beijing". This goal blends the concepts of green development, culture and intelligence.

8.5.1 Building Green Cities Across China

Green cities combine the concepts of low carbon, ecology, ecological civilization, circular economy, new energy sources, and so forth, with the aim to create green production methods and lifestyles and fully integrate the concept of ecological civilization into urban development. Globally, the primary experience in green urban development involves energy sources, transport, landscaping, industrial develop-

Table 8.4 Overview of main demonstration cities in China

Demonstration city name	Ministry/commission concerned	Implementation time	Main city or count
National Garden City	MOHURD	1992	16 batches, including 36 towns, 126 counties and 150 cities
National Forest City	State Forestry Administration (SFA) and National Afforestation and Greening Committee (NAGC)	2004	58 cities in 23 provinces, autonomous regions and municipalities directly under the Central Government
Pilot Low-carbon Province, Autonomous Region or Municipality	National Development and Reform Commission (NDRC)	August 2010	Batch 1 included 8 cities in 5 provinces; Batch 2 included 29 municipalities, autonomous regions and provinces
National Demonstration Circular Economy City or County	NDRC	December 2013	Batch 1 included 21 counties and 19 cities
Demonstration City Powered by New Energy Sources	National Energy Administration(NEA)	January 2014	Batch 1 included 81 cities and 8 industrial parks
National Demonstration Low-carbon City	MOHURD and Shenzhen Municipal Government	January 2010	Shenzhen as China's first ever National Demonstration Low-carbon Eco-city
National Demonstration City for Ecological Civilization	NDRC	December 2012	Guiyang as the first ever demonstration city of ecological civilization identified by the NDRC
National Pilot Demonstration Zone for Ecological Civilization	NDRC, Ministry of Finance (MOF), Ministry of Land and Resources (MLR), Ministry of Water Resources (MWR), Ministry of Agriculture (MOA) and SFA	December 2013	Fujian is China's first ever pilot demonstration zone for ecological civilization identified by the State Council
National Pilot Smart City	MOHURD	December 2012	Batch 1 included 90 cities; Batch 2 included 103 cities

(continued)

Table 8.4 (continued)

Demonstration city name	Ministry/commission concerned	Implementation time	Main city or count
National Demonstration Next-generation Internet City	NDRC, Ministry of Industry and Information Technology (MIIT), Ministry of Science and Technology (MOST) and State Administration of Press, Publication, Radio, Film and Television (SAPPRFT)	August 2013	Batch 1 included 22 cities

Source The author

ment, environmental protection and buildings (see Table 8.5). Based on this international experience, the Chinese government may promote green urban development across China by focusing on green industries, energy sources, transport, buildings, lifestyles and environmental governance.

1. **Encourage green industries**: It is necessary for the government to keep tight control over energy-consuming and heavily-polluting industries. Industrial parks should be improved so that more waste is recycled and the parks in themselves become more environmentally friendly. Greater efforts should be made to build demonstration parks with a circular economy and eco-friendly industries. Efficient and effective land use, waste exchange and reuse, cascading energy use, wastewater recycling and centralized pollutant treatment should be promoted in industrial parks.
2. **Optimize the energy mix**: It is necessary to accelerate the building of a renewable energy system, promote diverse, large-scale applications of distributed solar energy, wind power, biomass and geothermal energy and increase the percentages of new and renewable energy sources in the energy mix. The government should refer to the concept of the "Third Industrial Revolution" (Rifkin 2012) and promote the building of demonstration cities powered by new energy sources and the implementation of demonstration intelligent micro power grid projects. It is advisable to build distributed photovoltaic power generation demonstration zones in demonstration cities powered by new energy sources.
3. **Construct green buildings**: It is necessary to implement green building action plans, improve green building standards and certification systems, and expand the sphere of implementation. Measures should be taken to accelerate the improvement of existing buildings to make them more energy-efficient, to encourage the development and use of green building materials, and to promote housing industrialization. For example, existing residential buildings in northern China should be improved in terms of heat metering and energy conservation. Build-

Table 8.5 Green development measures in global cities

City	Main measure
Amsterdam	Environmentally friendly transport: 37% of its citizens travel by bike; entry of old vehicles into central urban areas is limited
Chicago	New energy sources: Hydrogen as fuel, wind power and solar energy Multi-tier landscaping: Tree planting and rooftop landscaping Green transport: Cycling
Curitiba	Unique public transport system Green industries: Factory building is prohibited in urban areas and suburbs
Freiburg	New energy sources: Solar power generation by solar panels at rooftops Green transport: One third of its citizens travel by bike MSW recycling
Kathmandu	Rooftop landscaping and building height limitations New energy sources: Solar energy
London	Green transport: Environmental tax for vehicles; bicycle rental service Energy-efficient systems
Reykjavik	New energy sources: Hydrogen buses and geothermal energy
Portland	Green buildings Green transport: Light rail, bus and bicycle; building green spaces and roads exclusively for walk or motor vehicles
Singapore	Zero-energy-consumption buildings: Power generation by rooftop solar panels, which connect with public power grids
Toronto	LED lighting systems, deep lake water cooling systems and rooftop landscaping

Source Relevant information

ings in areas where it is hot in summer and cold in winter, along with all public buildings, should be improved to become more energy-efficient. It is necessary to gradually increase the energy efficiency of new buildings in strict accordance with energy conservation standards and promote the industrialization and standardization of buildings so as to increase the percentage of residential buildings constructed and supplied in an industrialized manner.

4. **Encourage green transport**: It is necessary to accelerate the increased use of new energy vehicles and improve conditions for travel on foot or by bicycle in order to encourage green transport. Measures should be taken to accelerate the development of environmentally friendly vehicles powered by new energy sources. Supportive facilities such as charging stations/posts and gas filling stations should be built more rapidly, along with slow transport systems for pedestrians and cyclists. Effort should also be made to boost demonstration applications of vehicles in the public transport sector powered by new energy sources and cleaner fuel, including hybrid, all-electric vehicles and natural gas-powered

vehicles. It is necessary to continue the commercial vehicle use permit system based on fuel consumption and eliminate non-compliant vehicles more rapidly.

5. **Encourage green lifestyles**: It is necessary to accelerate the transformation of lifestyles toward simplicity, moderation, low carbon footprint and conservation in all aspects of daily life. Great effort should be made to develop an eco-friendly culture that encourages green consumption, the use of energy-efficient, environmentally-friendly vehicles and the construction of energy- and land-saving residential buildings. It is also necessary for cities to improve systems for recycling used items and turning leftovers into resources. Over-packaging and use of disposable items should be minimized.
6. **Enhance environmental governance**: Measures should be taken to efficiently and effectively use resources such as land, water and energy, promote recycling and reduce total consumption. Action plans for preventing and controlling air pollution should be implemented. For example, regional joint prevention and control may be carried out to improve air quality in cities. The systems for recycling used items and treating MSW on a per-category basis should be improved to enhance the reuse and treatment of MSW into harmless elements. It is necessary to define reasonable red lines for ecosystem protection, expand ecological spaces in cities, and build ecological corridors in urban areas.

8.5.2 Accelerating the Development of Cities of Culture

In the past, cultural inheritance was overlooked in the urban development process, resulting in damage and destruction of many historic sites and culturally significant buildings. Cultural inheritance is as important as the people-centric approach and ecological civilization in promoting new urbanization. Establishing cities of culture is an important approach to preserving cultural heritage in China.

Emphasizing cultural differences and individual city character is an important trend for urban development, especially at a time when cities are becoming increasingly similar. Nanfang Media Group (2011) assessed the cultural elements of Chinese cities in eight dimensions: livability, cultural environment, cultural heritage and growth, intelligentsia and cultural literacy, freedom of speech and power of the public voice, cultural character, cultural education and cultural events.

Taipei was identified as the paragon for cities of culture due to the prevalence of traditional culture, the right to free speech, numerous cultural landmarks and festivities, and the combination of internationalism with folk customs. The assessment from Nanfang Media Group noted that "[Taipei] has fascinating old streets and its people behave in a way reminiscent of the Republic of China and Confucianism." In comparison, the cultural aspect of Beijing is extremely controversial. Beijing is often described as a city with culture but without humaneness for damage to the traditional culture and buildings during urban development. Plus, problems such as traffic congestion and smog undermine Beijing's otherwise rich cultural heritage and high cultural creativity.

In order to build cities of culture, it is necessary to quickly establish national standards for planning and building such cities, and to encourage local governments to apply to build demonstration cities of culture. According to the experience of typical cities of culture, competent authorities should emphasize both the protection and integration of tangible cultural resources and the preservation of traditional culture through cultural inheritance. To build cities of culture in the future, the government should focus on promoting building and cultural integration and developing modern cities into culturally charming spaces with rich heritage and historic features.

It is necessary to value the protection and development of tangible cultural resources. inter-regional differences related to natural, historical and cultural endowments in each region should be embraced. Diversity should be formally articulated as a goal to prevent cities from copying one another. Great efforts should be made to develop beautiful towns and cities with historical memories, cultural heritage, locally unique landscapes and ethnic features. The preservation of historical and cultural heritage, ethnic cultures and traditional styles should be emphasized in the renewal of old urban areas. Functional improvement should be made in combination with the protection of culture and cultural relics. More effective measures should be taken to preserve and harness cultural resources and protect the cultural ecosystem in streets, towns and cities known for historical and cultural heritage as well as towns known for ethnic cultures. It is necessary to inherit and develop excellent traditional cultures, promote the development of locally distinctive cultures and maintain the cultural memories of cities.

It is also necessary to prioritize the inheritance and development of traditional cultures. In addition to tangible buildings of historical significance, the inheritance of traditional culture should be a core value when building cities of culture. It is necessary to develop and practice core socialist values, improve cultural management, production and operation systems and mechanisms more rapidly, and establish or improve the modern public cultural service system and modern cultural market system in accordance with the National New Urbanization Plan (2014–2020). Competent authorities should encourage urban cultures to diversify, as well as encourage the integration of traditional and modern cultures and local and non-local cultures, leading to a diversity of open, modern urban cultures.

8.5.3 *Building Smart Cities*

Smart cities arise from a combination of industrialization, informatization and urbanization. They are a higher form of urban informatization after digital and intelligent cities. The "smart city" concept was first proposed by IBM in 2008, and a wave of smart city programs swept the globe in 2010. For a city with a population of at least one million people, implementing comprehensive smart city management input will increase a city's development dividend by 250–300%, all other inputs remaining equal, according to estimation by the World Bank. This means that a smart city would be able to achieve about four times as much as original sustainable development-related targets and set the global trends in urban development (Gu et al. 2013). China has been rapidly building smart cities since 2010, following the emergence of new-generation IT such as the Internet of things (IoT) and cloud computing, As of August 2013, there were already 193 pilot national smart cities in areas including Wuxi, Pudong New Area, Ningbo, Shanghai, Hangzhou, Beijing, Guangzhou, Foshan and Xiamen. These are the leading areas in China in terms of smart industries, infrastructures and applications.

Yet fast development has caused significant problems regarding top-down design, core technology and specifications/standards for smart cities in China. Gu et al. (2013) have pointed out that several problems now affect the building of smart cities in China, including the prioritization of projects over plans, construction over application and imitation over R&D. There is also a great deal of difficulty in consolidating resources. Gu et al. (2013) summarizes these difficulties into three main challenges. First, a national plan for developing smart cities has yet to be made. The building of smart cities in China is always connected with the industrial development of new-generation IT such as the IoT and cloud computing. Given the lack of a national plan for building smart cities, there are big differences between local governments in the understanding, planning and building of such cities. Second, there is a shortage of locally-developed core technologies. Given the lack of locally developed core technologies in the smart industry supply chain, such as high-end chips, key components and underlying software, China has to import most of the core technologies and products for building smart cities. Such reliance on foreign countries is detrimental to the healthy growth of relevant industries and will cause potential information security problems. Third, relevant standards and specifications are not yet uniform. The building of smart cities covers a variety of fields, but there is no single set of industry, construction and assessment standards. Complex interfaces between different systems make it difficult to interconnect systems and share and coordinate information.

In order to promote the building of smart cities in the future, the focus should be on top-down design, standards development and technological R&D. To enhance top-down planning and design, it is necessary to more rapidly complete a national plan for building smart cities, clarify the mid- and long-term targets and main tasks for this process, and guide local governments, enterprises and other organizations in systematically building smart cities under the *National New Urbanization Plan (2014–2020)*. Second, the State should more rapidly promulgate administrative laws and regulations regarding the building of smart cities, and relevant ministries/commissions should establish corresponding systems and specifications. It is necessary to more rapidly improve and implement relevant technologies, applications and management standards in fields such as IoT, cloud computing, communications and networking. Smart city assessment metrics should be created. Third, it is necessary to increase investment in technological innovation, make breakthroughs in core technologies such as high-end chips, key components and underlying software, and enhance technical support systems. This will reduce reliance on imported technologies and products. Great efforts should be made to lay a solid basis for information security so as to ensure that the entirety of information security in Chinese cities is under control. The Chinese government should provide strategical guidance and establish relevant support systems to gradually promote the building of comprehensive smart cities with a full range of smart application systems covering smart public services, urban complexes, integrated urban management and operation platforms, resident services, education and cultural services, service applications, health assurance systems, transport, and so forth.

References

Chi, Z. R., Wang, H.C., & Chen, Y. (2008). On the grid-based urban management model in Shanghai. *Science & Technology Progress and Policy, 1.*

Ding, Z. Y. (2013). On inter-city differences in the ratio of housing price to income in China. *Journal of East China Normal University (Philosophy and Social Sciences), 3.*

Feng, H. (2014). On the development of villages in the city during urbanization. *China Construction, 2.*

Gu, S. Z., Yang, J. W., & Liu, J. R. (2013). The development of smart cities in China: problems and countermeasures. *China Soft Science, 1.*

Jiang, A. L., & Ren, Z. R. (2007). Grid: a new urban management model. *Shanghai Urban Planning Review, 1.*

Li, T. (2013). *Urbanization: Comprehensive and profound social transformation.* Beijing: China Development Press.

Liu, S. Y. (2009). On the ways of public participation in urban management in China. *Journal of Tongji University (Social Science Section), 3.*

Nanfang Media Group. The Top 10 cities of culture. (2011). *Southern Weekly*, December 14, 2011.

Qiu, B. X. (2012). A new type of urbanization: from concept to action. *Administration Reform, 11.*

Rifkin, J. (2012). *The third industrial revolution.* Beijing: China CITIC Press.

Song, G., & Tang, Q. (2007). On modern cities and their management—A type of open, complex, gigantic systems. *Urban Studies, 2.*

Sun, Y. L., Liu, H., Shen, S. Q., & Lin, L. L. (2012). Analysis on flooding in Chinese cities: Based on surveys in Wuhan, Shanghai, Hangzhou and Beijing. *JINTIAN, 8.*

Tao, R. (2012). Thoughts on how to accelerate the renewal of villages in the city and promote the building of affordable housing. *Administration Reform, 1.*

Wang, D. W., Wang, Y. C., & Su, Y. (2012). How serious are urban diseases in China: The measurement of urban diseases and quantitative comparison between cities. *China Development Observation, 10.*

Xiang, G. Q. (2005). On developed countries' experience in renewing old urban areas and its inspiration for urban redevelopment in China. *Academia Bimestris, 4.*

Yuan, D. Z. (2005). How do foreign countries deal with urban diseases? *Scientific Decision Making, 8.*

Yang, M. S. (2013). Short-lived buildings are an urban disease and power-related disease. *China Construction News*, November 22, 2013.

Zhang, D. H. (2010). On shanty town redevelopment by local governments: Patterns and strategies. *Journal of Fujian Administration Institute, 1.*

Zhang, J. T. (2009). China must refer to international experience in renewing old urban areas. *Reform, 10.*

Zhai, B. H., Li, C., & Yang, F. (2011). Comprehensive modern urban management: A review of its essence, functions and systems. *Shanghai Urban Management, 6.*

Zhaopin.com and the Peking University Center of Social Survey Research. (2012). Research report on the balance index of Chinese employees 2012, http://wh.house.sina.com.cn/2012-04-26/1720130761.shtml.

Zhou, G. Z. (2002). Cities and their areas—A typical open, complex, gigantic system. *Urban Studies, 1*.

Chapter 9
Pushing Institutional Innovation in Urbanization

Urbanization is a complex, multifaceted process involving urban and rural areas, agriculture and non-agricultural industries, and urban and rural residents. Conducting urbanization requires top-down design and an overall framework for urban and rural integration. In order to conduct new urbanization with Chinese characteristics, the Chinese government must eliminate the hindrance of the dual urban-rural structure, accelerate integrated urban and rural reforms, and establish and improve a single set of household registration, land management, employment management, social security and administrative area management systems for urban and rural areas. This will facilitate the free flow of factors of production and fair exchange between urban and rural areas, optimize the allocation of public resources, and enhance coordination between population and land urbanization and between urbanization and hukou urbanization, thus improving the quality of urbanization across the board.

9.1 Accelerating Household Registration System Reform

Reform of China's household registration system is progressing far slower than desired, while urbanization and hukou urbanization are carried out as entirely separate processes. This has become the primary obstacle to improving the quality of urbanization. China's current *Regulations on Household Registration* was promulgated and took effect in 1985. It is now largely inconsistent with China's current situation and, hence, no longer suits social management requirements. In February 2011, the General Office of the State Council released the *Notice on Pushing Household Registration System Reform in an Active and Prudent Manner*. This Notice required that the household registration transfer policy be determined on a per-category basis. Moreover, having a legal, stable job and domicile, including rented dwellings, would now be a basic condition for becoming registered on the *hukou* system in one's place of residence. Both the 18th CPC National Congress and its 3rd Plenary Session required that the household registration system undergo more rapid reforms.

H. Wei, *Urbanization in China*, Research Series on the Chinese Dream and China's Development Path, https://doi.org/10.1007/978-981-13-1408-7_9

The 2014 Central Urbanization Work Conference called for local governments to resolve the *hukou* issue for rural people who have already migrated to towns and cities for work. Under the guidance of the CPC Central Committee and the State Council, local governments have successively eliminated the division between rural and urban *hukou* in recent years and loosened the household registration policy for rural migrants living in cities. Chinese towns and small/medium-sized cities, in particular, have generally lifted limitations on such registration, while some larger cities have also lowered the barriers to registration. Specifically, Guangdong, Shanghai and Tianjin have established a points-based household registration system; Chongqing, Chengdu and Suzhou have implemented a policy that allows locals to freely transfer their *hukou*. Most provinces have established a residence permit system and a single household registration system for urban and rural areas. Between 2010 and 2012, a total of 25.05 million people across China had their rural *hukou* changed to an urban *hukou*. This number was 2.2 times as high as in the previous three-year period.

From an overall perspective, however, reform of the current household registration system is still limited to lifting limitations on *hukou* transfer. Reform of the social welfare system has yet to occur. Local social security systems are still based upon the household registration system. Schools in some places still require a local *hukou* as a precondition for admission. Policies are still in place that discriminate against people with rural *hukou* in terms of access to birth control, post-retirement arrangements for conscript soldiers, preferential treatment and arrangements for entitled groups and compensation for death in traffic accidents. Many larger cities have set restrictions on vehicle and housing purchases by non-locals and some even have *hukou* requirements for employment. Some very large and ultra-large cities still set household registration quotas for non-local graduates and assess non-local graduates to see whether they meet certain requirements. Some local governments have eliminated the division between rural and urban *hukou*, but the rights and benefits available depending upon the type of *hukou* have not been uniform. In other words, reform of the household registration system intended to unify urban and rural *hukou* has yet to make significant progress. With regard to the implementation of local points-based household registration systems, the number of people who have successfully registered on the host city's *hukou* system in recent years is still very small due to unreasonably high barriers and tight quotas. The residence permit system has partially solved problems facing rural migrants in terms of citizen's treatment, but this system does not yet cover benefits such as the minimum living allowance for urban residents and public rental housing support. Since the *Residence Permit Management Measures* has not been issued, local governments vary from each other in terms of fee standards, applicability, holders' rights and obligations, and the number of years after which holders can get registered on the local household registration system. Accordingly, household registration system reform must be accelerated in order to fuel a new type of people-oriented urbanization, eliminate the dual structures between urban and rural areas and within cities, and improve the quality of urbanization.

9.1.1 Determining the Basic Direction of Household Registration System Reform

The household registration system is a national administrative system for legally gathering, verifying and providing basic household information. Its basic functions include proving identity, collecting population statistics and conducting social management. Reform of the household registration system is aimed at eliminating the urban-rural divide, registering residents in their permanent places of residence, implementing a uniform household registration and management system for urban and rural areas, and most importantly, making rights and benefits independent of *hukou* type in order to achieve equality in citizen status and rights. Since household registration reform involves various aspects and receives wide-spread attention from across society, it is critical that the Chinese government develop a good overall plan and top-down design to carry out voluntary, categorical and systematic reform in an active yet prudent manner. The current practice in which local governments direct household registration system reform should be changed to allow the Central Government to play the leading role. An overall program for nationwide reform of the household registration system should be made at the national level and released as soon as possible. This national program should spell out ultimate objectives, institutional configurations, a roadmap for reform, and an implementation schedule. Competent authorities have announced that a new, uniform, people-oriented, efficient household registration system for urban and rural areas will be established by the end of 2020. Under this new system, the basic condition for transferring one's *hukou* is possession of a legal, stable job and domicile in the permanent place of residence to which the *hukou* is intended to be transferred. Yet the requirements under this new system are no different than the current practice and do not address the ultimate goal of household registration system reform, to establish a uniform household registration and management system for urban and rural areas, implemented under one sole criterion—registering residents' information in their permanent places of residence—with no other conditions such as employment and domicile.

9.1.2 Delinking Benefits Distribution from Household Registration Category

Successful household registration system reform depends on making benefits distribution independent of *hukou* type. At present, there are more than 20 rights and benefits tied to *hukou*, including democratic rights, employment opportunities, childhood education, social security, access to birth control, vehicle and housing purchases, post-retirement job arrangements for conscript soldiers, compensation for death and injury in traffic accidents, subsidies/allowances, and so forth. In order to accelerate household registration system reform, the Chinese government must change all policies tied to *hukou* type and make rights and benefits independent of *hukou* as early as possible. Policy reform should cover the entire permanent population, not just the urban permanent population, with basic public services offered to all registered permanent residents. All citizens should be eligible for basic civil rights, including the right to vote, stand for election and receive public benefits. All citizens should also be able to receive public services and benefits in their permanent place of residence, namely, wherever their *hukou* is registered. New local policies should be prohibited from being tied to *hukou* type. *Hukou*-independent policies should extend to social security and public services fields, in addition to employment, compulsory education and skills training already specified by the State Council. New population control policies should also not be tied to *hukou* type, even in ultra-large cities such as Beijing and Shanghai. Non-discriminatory criteria should be applied instead. Moreover, *hukou*-dependent policies regarding employment, education, birth control, medical care, elderly care, housing and others should be changed. Discriminatory criteria set according to *hukou* type also should be canceled and replaced with new, uniform criteria for urban and rural areas, before gradually making these policies independent of *hukou* altogether. Relevant policies should be adjusted as soon as possible and relevant laws and regulations be amended wherever possible. Otherwise, methods for step-by-step implementation should be developed and all policies disconnected from *hukou* according to a set implementation schedule. These measures will lead to a uniform social security system for urban and rural areas and a mechanism for equal access to basic public services.

9.1.3 Establishing a Uniform Residence Permit System in China

The Chinese government should release the *Residence Permit Management Measures* as soon as possible in order to improve the residence permit system. Key changes to the residence permit system should include: the gradual removal of restrictions for residence permit applications, including the removal of preconditions related to educational background, professional title, criminal record, and so forth.; the requirement that all local governments must grant residence permits to permanent residents with non-local *hukou* without any limitations; and equal treatment for both holders of residence permits and residents with local *hukou* with respect to the right to vote, the right to employment, compulsory education, skills training, temporary assistance, basic medical insurance, basic pension insurance and unemployment insurance, and so forth., in order to protect basic civil rights and benefits. Social security and public services other than basic security (i.e. housing support and general social assistance) should be determined by local governments, not the Central Government, depending on the specific situation. On this basis, local governments may establish a points-based *hukou* system tied to conditions such as the number of years of residence in the host city, job or business status, stability in source of income, and social security and tax payments. Residence permit holders should be eligible to receive a local *hukou* if they meet a certain points level. The requirements to receive a local *hukou* should not be unduly burdensome, and since residence permits are a transitional measure, the transition period to a local *hukou* should not be too long. Residence permits should be canceled within a certain length of time, and all holders of residence permits should then be registered on the *hukou* system in their permanent place of residence, thus replacing the residence permit system with a uniform *hukou* system across China. The "right time" to transition to a uniform *hukou* system is when mechanisms are in place to provide China's entire permanent population with equal access to basic public services. Given the current situation, it may take 10 or more years to replace the residence permit system with a uniform *hukou* system across China.

9.1.4 Reforming the Household Registration System in a Localized, Phased Manner

Since reform of the household registration system is a complex, long-term task, short term efforts should be coordinated with long-term efforts to steadily push forward reform in a phase by phase manner, with specific objectives, tasks and concrete measures implemented in each phase. From a national perspective, the household registration system should be reformed according to a two-pronged approach, adjusting *hukou* policies in the short term while pursuing real reform to the household registration system in the long-term. With respect to the later, a uniform household registration and management system, a single set of systems/mechanisms for urban and rural areas and a mechanism for equal access to basic public services should be implemented by 2025. This two-pronged approach will allow municipal governments to continue the current practice of loosening requirements for granting local *hukou*, starting with fewer restrictions for rural migrant workers who are already permanent residents with legal and stable jobs and domiciles. At the same time, it is necessary to disconnect benefits from *hukou* and gradually establish a mechanism for equal access to basic public services and a larger, single set of social security, employment management, land management and social management systems for urban and rural areas. The Chinese government should provide the entire permanent population with basic public services according to the permanent population registered. The success of this reform depends upon establishing a mechanism for equal access to basic public services and a larger, single set of systems/mechanisms for urban and rural areas. Once this set of systems and mechanisms are in place, the current reformative practice of loosening local *hukou* requirements will become obsolete.

Keeping in mind overall carrying capacity and growth potential, Chinese cities should implement practical, specific programs to reform the household registration system on a per-category and per-step basis in line with national reform objectives. The 3rd Plenary Session of the 18th CPC Central Committee and the Central Urbanization Work Conference determined that "The country will relax overall control of farmers settling in towns and small cities, and relax restrictions on settling in medium-sized cities in an orderly manner. China should set reasonable requirements for rural residents to obtain *hukou* in large cities, and strictly control the size of the population in mega-cities." This means that the household registration system reform will proceed more rapidly in smaller cities with fewer difficulties in this regard. Towns and small/medium-sized cities are urged to grant urban *hukou* to permanent residents with rural *hukou* on a voluntary basis as soon as possible, to enable full and equal treatment for permanent residents. Household registration reform will occur the most slowly in large cities and mega-cities given the large number of rural migrants and the attendant heavy pressure on hukou urbanization. These cities should now focus on addressing the *hukou* issue for rural migrant workers with stable jobs, those who have moved into these cities with their families, and new-generation rural migrant workers.

9.2 Improving the Urban and Rural Land Management Systems

China has long utilized separate urban and rural land management systems. Land in urban areas is owned by the State and is centrally managed by either the State or the city. In more recent years, some local governments have adopted a uniform approach to planning, expropriating, reserving, developing, transferring and managing land within planned urban areas. In comparison, land in rural and suburban areas is collectively-owned, managed by collective economic organizations or villagers' committees comprised of farmers. These separate urban and rural land management systems have caused separate urban and rural land markets and land administration, which in turn have resulted in numerous social conflicts during urbanization and inefficient use of land resources. Land for rural development cannot be bought and sold on the market, and can only turn into land for urban development after it is expropriated by the government. This system is prone to creating social conflicts. Historically low compensation for expropriated land has damaged farmers' interests and pushed urban sprawl, causing inefficient use of urban land. At the same time, the imperfect rural land market has also caused rural land to be left idle after farmers move into cities for work, and prevents farmers from capitalizing on rural land in cities. The Chinese government must eliminate institutional barriers regarding land and regulate land management related to land registration and rights verification, land market improvement, land expropriation and urban land use.

9.2.1 Establishing a Uniform Land Registration System

A uniform urban and rural land management system involves conducting uniform land rights registration and legally determining the owners of land use rights and the purposes of the used land. Under the current *Land Administration Law*, the rights to collectively-owned rural land are verified, registered and certified by county governments; the rights to state-owned urban land, by governments above the county level. For state-owned land used by central government departments/agencies, the State Council identifies the authorities for registration and certification. At the same time, the governing laws and authorities for rights verification and certification vary with the features permanently and immovably attached to land. For example, the certificate of rights for forested areas is issued by the forestry authority under the *Forestry Law*; the certificate of rights to grassland, by the agricultural authority under the *Grassland Law*; the certificate of rights to aquaculture, by the aquacultural authority under the *Fisheries Law*. This practice of separate administration is inconsistent with the *Land Administration Law*, which provides that "The land administration department under the State Council shall be in charge of uniform administration of land supervision over the land throughout the country." Moreover, this practice has created uncertainty and constraints with respect to the responsibilities of relevant government departments and agencies, and has created confusion and inconvenience to urban and rural residents applying for these certificates. It is therefore necessary to amend the *Land Administration Law* and other relevant laws to establish a uniform urban and rural land registration system as soon as possible. Amendments to these laws should clarify that the land administration is the sole authority in charge of verifying, registering and certifying the rights to urban or rural land. Second, amendments should unify land registration standards. All land—state-owned urban land, collectively-owned rural land, farmland, woodland, grassland or land not for rural development—should be covered by a uniform land registration system. In addition, uniform land registers and rights certificates should be issued and a uniform urban and rural land registration information system be established. Third and last, amendments should establish a uniform, land-centric real estate registration system for real estate in all sectors, including land, housing, grassland, woodland and sea areas.

9.2.2 Establishing a Uniform Urban and Rural Land Market

To deepen rural land system reform, it is necessary to rapidly establish a uniform urban and rural land market, especially for land used for urban and rural development. First, the Chinese government should create a uniform land system for urban and rural development. Relevant laws such as the *Land Administration Law*, the *Property Law*, the *Guaranty Law* and the *Urban Real Estate Administration Law* should be amended as soon as possible to remove provisions blocking the trade of rural, collectively-owned buildable land under the same terms and conditions as state-owned land. The laws should also be amended to establish a uniform, regulated land system for urban and rural development. Article 10 of the *Constitution*—"Land in the cities is owned by the State"—also should be amended according to the new situation. Second, the Chinese government should adhere to the principle of trading under equal terms and conditions and the same rights and prices for both rural, collectively-owned buildable land (mostly held by TVEs) and state-owned land. The transfer, leasing and mortgaging of rural land in both the primary and secondary markets should be allowed (Jiang 2013). The buying and selling of rural, collectively owned buildable land should be allowed under the same terms and conditions as for state-owned land, under the premise that it conforms to planning. The building and trading of rural houses with limited property rights, for example, are not protected by law because such houses do not conform to planning. It is therefore necessary to make policies and measures specific to such houses as soon as possible. Third, the Chinese government should regulate the transfer of rural, collectively owned buildable land. Based on the experiences of provinces including Guangdong and Zhejiang, the Chinese government should expeditiously enact *Regulations on the Transfer of Rural, Collectively Owned Buildable Land* in order to unify provisions regarding legal duties and preconditions, means and procedures for trading and profiting from land. Rural, collectively owned buildable land should be traded via invitations to bid, auctions and listings on a uniform platform. Fourth, the Chinese government should ensure farmers' usufruct of homestead. The 3rd Plenary Session of the 18th CPC Central Committee made clear that the Chinese government "will … improve the rural homestead system through reform. We will select several pilot areas to steadily and prudently push forward the mortgage, guarantee and transfer of farmers' residential property rights, and expand the channels for farmers to increase their property income." Homestead land falls under non-commercial buildable land, and is not yet eligible for trade on the same terms and conditions as state-owned land. From a long-term perspective, however, pilot projects should be launched to explore, including homestead land into the uniform market for land used for urban and rural development. Fifth and last, the Chinese government should build a uniform land trading platform for urban and rural areas. This platform and related trading procedures should cover rights verification, registration and certification, farmer-contracted land, collectively owned buildable land, homestead land, the right to use woodland, and so forth. and be standardized to promote the reasonable, orderly transfer of rural, collective owned land.

9.2.3 Improving Policies for Expropriating Rural Land

Throughout the process of urbanization, the Chinese government has compensated farmers for expropriated rural land at a rate based on the land's agricultural purpose. Specifically, both the compensation for land and subsidy for resettlement equal an amount slightly greater than the annual average output value of the rural land over the three years prior to the land's expropriation. This has resulted in extremely low compensation to farmers. The government typically transfers the expropriated land at a much higher price based on the market value of land used for urban development. Low-cost expropriation combined with high-price transfer has become an important means for local governments to increase revenue. Low compensation for expropriated land, plus an unreasonably wide range of land that can be expropriated, has caused a rapid increase in urban land, a large number of farmers who have lost their land,

and a large number of disputes regarding land expropriation. Consequently, there is a great deal of farmer complaints and protests occur frequently. To address this issue, it is necessary to narrow the range of land that can be expropriated, standardize the procedures for land expropriation, and change the method of compensation based on agricultural output value. Compensation should instead be paid according to market value. Additionally, a reasonable, standardized, diversified set of support mechanisms should be established for farmers whose land has been expropriated to enable farmers to share in the benefits of urbanization.

With respect to narrowing the range of land that can be expropriated, the Chinese government must first clearly define land used for the public interest and land for commercial purposes. Expropriated rural land should be limited to land used for the public interest, thereby gradually reducing the amount of rural land expropriated in a forcible manner. The government must not forcibly expropriate land for commercial purposes, and instead should focus on improving the market for rural, collectively-owned buildable land. With respect to standardizing land expropriation procedures, the Chinese government should improve procedures relating to expropriation approval, implementation, resettlement and dispute investigation/arbitration. The Chinese government should also take measures to ensure farmers' rights to information, participation, monitoring, and filing complaints. The government should also protect the right of farmers to profit from expropriated land, and regulate and restrict land expropriation by local governments to prevent abuse of power (Jiang 2013). With respect to improving the support mechanism for farmers, it is necessary to increase the rate of compensation for expropriated land to be based on market value rather than agricultural output value. The rate disclosure mechanism also should be improved to ensure that land of equal condition is compensated at equal rates, thus effectively protecting farmers' legal profits. Comprehensive policies and measures regarding employment support, vocational training, housing support, social security and housing resettlement should be adopted to ensure the rights, benefits and long-term livelihood of farmers whose land has been expropriated.

9.2.4 Implementing Uniform Urban and Rural Land Administration

From a long-term perspective, new urbanization requires uniform urban and rural land administration, which can be accomplished by establishing a uniform urban and rural land management system. Up to now, urban and rural land in China has been managed in different manners. Rural land is managed with a focus on land use, while urban land is managed with a focus on ownership (Wu and Wu 2009). Cadastral management remains confined to urban areas, while rural land is still managed mainly in terms of land use, land distribution, and changes to land use. Rural cadastral surveys began in China only a few years ago. To implement uniform urban and land administration several measures are needed. First, the Chinese government must enhance cadastral surveys. Such surveys should be made in combination with land surveys and conducted under the principles of lawful ownership, clear boundaries and accuracy. These surveys should provide comprehensive information on every piece of urban or rural land in terms of use, area, ownership, boundaries, and so forth., with particular attention paid to ownership of rural, collectively-owned land and land use rights for construction. Second, the Chinese government should build a uniform cadastral information center. Based on the results of the cadastral surveys, the Chinese government should consolidate relevant departmental resources and data on urban and rural land to build a uniform, nationally networked, urban and rural cadastral information center open to the entire society. This will facilitate resources sharing. Third, the Chinese government should build a uniform land statistics system. The land administration and statistics authority should be responsible for building a single set of land categorization, statistics and indicator systems for urban and rural areas and releasing land data in a centralized manner. When necessary, the land administration and statistics authority should address instances where the statistics may be false or inaccurate. Fourth and last, the Chinese government should conduct uniform urban and rural land administration. The power of land administration should remain centralized and, more importantly, there should not be separate administration systems based on urban and rural differences, the nature of land ownership or features permanently and immovably attached to the land. A single land authority should conduct urban and rural land administration in a uniform manner under law. Land use permits for state-owned forests, grassland and agricultural land should only be granted by the land authority on behalf of the State Council, with differences in land use permitted.

9.3 Improving Urban and Rural Employment Management Systems

Improving the urban and rural labor markets and employment management system is critical for the hukou urbanization of rural migrants and increasing the quality of population urbanization. In March 2006, the outline of the 11th Five-Year Plan stated that urban and rural employment should be managed in a uniform manner. In July 2006, four ministries and commissions including the then Ministry of Labor and Social Security (MLSS) co-issued the *Guiding Opinion on Pilot Projects for Uniform Urban and Rural Employment Management*. This guiding opinion launched pilot projects across China with an aim to establish an integrated urban and rural labor market, develop institutional configurations for equal access to employment opportunities for rural and urban residents, and promote sufficient employment. In October 2008, the 3rd Plenary Session of the 17th CPC Central Committee announced the goal of managing urban and rural employment in a uniform manner and rapidly establishing an integrated urban and rural human resource market. In November 2013, the 3rd Plenary Session of the 18th CPC Central Committee noted that "We will regulate human resource management, removing all institutional barriers and employment discrimination that affect equal employment, such as the birthplace of a person (cities or countryside), industries, social status and gender … We will improve the public service system for employment that favors both urban and rural areas equally." Under this uniform deployment by the Central Government, local governments have been working to integrate urban and rural employment and have made remarkable progress in recent years. Employment discrimination against rural migrant workers is gradually diminishing. Nonetheless, separate urban and rural labor markets remain in China and urban and rural employment inequality is still obvious. It is therefore necessary to continue reforming the employment system so as to facilitate the free flow of workers between urban and rural areas. Measures should be taken to eradicate employment limitations and discrimination against rural migrant workers. In order to promote equal and sufficient employment of urban and rural residents, it is necessary to establish a uniform urban and rural employment/unemployment registration system and a public employment and business startup service system equally accessible to all. Efforts should be made to establish a well-regulated, integrated urban and rural labor market characterized by fair competition.

9.3.1 Removing All Barriers to Equal Employment

Equal employment refers to equal rights and opportunities for employment for both urban and rural residents. Separate urban and rural *hukou* systems and labor markets have long caused severe employment discrimination against rural migrant workers. There are two main aspects to this discrimination. The first aspect is unequal employment opportunities. To date, some local governments still have entry limitations against rural migrant workers in certain industries and jobs. For example, some local governments explicitly prohibit or restrict rural migrant workers from taking jobs such as civil servants, employees at public institutions, bus/taxi drivers and bank employees. Local *hukou* is required for these particular jobs. As another example, public employment services in some places still differentiate employee candidates based on *hukou* and region of origin—they prioritize the provision of employment, business startups and training services to locals, and may even refuse to serve non-local rural migrant workers (Wang 2013). The second aspect is unequal treatment. Rural migrant workers face discrimination and difficulties in receiving payment for work, getting days off for rest and holidays, and receiving protection at work. They are also frequently subject to forced labor and receive lower employment subsidies and lower social security benefits. Compared with urban employees at state-owned organizations, rural migrant workers generally face longer work hours, lower salaries, lack of benefits and unassured personal rights. In 2012, 41.3% of all employers of rural migrant workers did not provide accommodation or housing subsidies (the NBS 2013). Rural migrant workers also lagged significantly behind urban employees in terms of wages and participation in urban social security programs.

It is generally believed that the separate urban and rural *hukou* systems are the primary cause for employment discrimination against rural migrant workers. Scholars have found that the *hukou* system is a major factor. Urban-rural differences in *hukou* represent 76% of the wage gap between urban and rural workers, according to a study by Cai et al. (2003). A study by Hu (2007) finds that the *hukou* system contributes 66.9% to the aforementioned gap. Yet, as reforms continue, the *hukou* factor is playing less of a role in this gap. While it is necessary to continue reforming the *hukou* system, it is just as important to improve relevant systems, mechanisms and policies to eradicate all barriers to equal employment, particularly the existing employment limitations against rural migrant workers. All organizations should recruit urban and rural workers in a uniform manner. All forms of employment discrimination against rural migrant workers must be prohibited. Rural migrant workers should receive assured employment-relevant rights and benefits identical to the rights available to urban people. Legal measures should be put in place to ensure that rural migrant workers receive the same pay and same benefits for the same job. Lastly, it is necessary to establish an integrated employment policy system that provides equal services and support to urban and rural people, including those related to employment opportunities, business startup support, worker protection, employment management, and so forth.

9.3.2 Establishing a Uniform Urban and Rural Employment/Unemployment Registration System

China has had an urban employment/unemployment registration system since the early 1980s, yet there is no such system available to the rural workforce. Moreover, governmental employment assistance programs and service policies only focus on urban workers. By excluding rural workers who are available and want to work, the separate employment management systems discriminate against the rural workforce while at the same time inaccurately reflect the status of employment and unemployment across China. These separate systems are therefore detrimental to macro regulation and social stability. To date, relevant government departments still only release data on the number of urban employees and the registered urban unemployment rate. This unemployment rate tends to be much lower than reality and does not fully reflect the overall status of unemployment across China since it excludes rural migrant workers and jobless urban residents not willing to register their unemployed status (Yang and Gu 2011; Cai and Wang 2009). It is therefore necessary to expeditiously establish and implement a uniform urban and rural employment/unemployment registration system across China. The experiences of Jiangsu Province, Qingdao City, Shandong Province and Chengdu City, Sichuan Province should serve as models for such implementation. Specific measures should include providing uniform employment/unemployment registration for the rural workforce, issuing uniform certificates once workers register for employment/unemployment, and regularly releasing the registered urban and rural unemployment rate. In Chengdu City, for example, the municipal unemployment registration and management services bureau already covers jobless urban and rural residents who are above the legal working age, available for work and want to work, who have no right to contract land or use woodland in the countryside or have transferred such rights. This experience is worth spreading across China. At the right time, the registered unemployment rate should be replaced with an unemployment rate based on surveys.

9.3.3 Building a Public Employment and Business Startup Service System Equally Accessible to Urban and Rural Residents

Public employment and business startup services refer to non-commercial employment and business startup services provided by governments at all levels. They are an important part of basic public services in China. These services aim to provide equal employment and business startup opportunities to people who are available and willing to work or start up their own business. China's policies regarding public employment and business startup services have long been biased toward cities. Chinese government employment targets focus on lowering the registered urban unemployment rate and fund/policy support is centered on the reemployment of jobless urban residents. In stark contrast, the Chinese government does not yet monitor the status of employment/unemployment of rural residents, nor does it cover rural workers under its macroeconomic management target system or offer sufficient financial inputs and incentives. This has led to inequality between urban and rural residents in terms of employment opportunities. Local governments have started to provide the rural workforce with public employment and business startup services in recent years, but they have yet to overcome the challenges caused by the inequality between urban and rural residents. Local governments should make uniform arrangements for urban and rural employment and business startup services based on the principles of urban and rural integration, impartiality and fairness. They should start by building a uniform public service platform to provide public employment and business startup services equally accessible to urban and rural residents. To do so, they must first consolidate the separate labor markets, including the labor service and human resource markets, into a uniform labor market that covers urban and rural areas as well as all industries and sectors. This is an important step to building a uniform, fair, open and standardized public employment service system. Second, local governments should provide public employment and business startup services—such as job seeker registration, job recommendations, employment guidance, vocational training and business startup support- to both urban and rural residents on an equal basis and under a single set of standards. It is necessary to increase financial inputs, enhance vocational training for rural migrants and ensure that rural migrants are covered under the national education and training system. Third, local governments should provide employment support and assistance services to farmers who have lost their farmland and to jobless rural residents who are having difficulty finding a job. It is important that employment assistance policies -such as social insurance subsidies, unemployment insurance benefits, startup subsidies, and identification services for individuals having difficulty finding a job-seamlessly cover all urban and rural areas.

9.4 Improving Urban and Rural Social Security Systems

As social security system reform continues in China, basic pension and medical insurance programs have extended to cover a greater number of both urban and rural residents in recent years. A basic social security framework that focuses on pension and medical insurance and basic living allowances has been put in place (Hu 2012). China has now entered a new stage of full social security program coverage. Nevertheless, social security in China still faces numerous problems, including a low level of uniform arrangements, an imbalance between urban and rural development, an imperfect inter-connectivity mechanism, and low levels of social security benefits in rural areas. The perennial division between urban and rural social security systems still persists. In 2009, urban residents accounted for over 90% of the social security benefits offered by the Chinese government, indicating that social security benefits available to rural residents are insufficient (Wang 2012). At the same time, the separate urban and rural social security systems have posed a significant challenge to social security programs targeting people moving between urban and rural areas, particularly rural migrants. This challenge is exacerbated by the lack of a reasonable inter-connectivity mechanism. Removing the urban-rural divide by establishing a uniform set of basic social security systems for urban and rural residents is a common practice in developed countries and is a necessary precondition for building a moderately prosperous society and promoting socialism with Chinese characteristics. In order to achieve new urbanization, the Chinese government should work to improve relevant mechanisms and increase social security coverage with a focus on increasing fairness and adapting to mobility. It should do so by adhering to principles of full coverage, expanding the basic safety net at multiple levels, benefits shared by all, urban and rural integration, equal services, and sustainability. In this way, the Chinese government can achieve the ultimate goal of making basic social security benefits available to all Chinese people.

9.4.1 Building a Uniform Basic Medical Insurance System for Urban and Rural Residents

There are three basic medical insurance systems in China, the medical insurance system for urban employees, the medical insurance system for urban residents and the new rural cooperative medical care system (NRCMCS) (see Table 9.1). Implemented across China in late 1998, the basic medical insurance system for urban employees is compulsory in nature and targets all urban employees. It covered 265 million people in 2012. Built upon pilot projects, the NRCMCS was established in 2013 and targets rural residents. It covered 832 million people in 2011, suggesting a coverage rate of 97.5%. The basic medical insurance system for urban residents was set up in 2007 and mainly targets elementary, middle and high school students as well as other non-working residents in urban areas. It covered 272 million people in 2012. Today, China's medical insurance systems theoretically cover China's entire population. That said, full coverage does not mean that all urban and rural residents are covered by medical insurance. The three basic medical insurance systems are fragmented, target separate groups in urban and rural areas, and differ from one another in terms of management, fundraising, level of benefits, operating mode and level of reimbursement. There is also a complete lack of inter-connectivity between urban and rural systems. The basic medical insurance systems for urban employees and urban residents are managed by the Chinese government's human resource and social security authorities, while the NRCMCS is managed by the health and birth control authorities. Such fragmentation has led to inequality in provision of services between urban and rural residents and between different groups. There are also problems related to overlapping coverage, financial waste, and mutual management constraints. It is therefore necessary to further consolidate resources, make uniform arrangements for urban and rural medical insurance services, and build a uniform basic medical insurance system for urban and rural areas. This will enable equal basic medical insurance services for both rural and urban residents.

In May 2013, the NDRC issued the *Opinion on Main Tasks for Continuing Economic Reform in 2013*, which established the aim to consolidate managerial functions related to urban and rural basic medical insurance systems, gradually unify the basic medical insurance systems for urban and rural residents, and establish a sound medical insurance system for all people. The *Decision on Major Issues Concerning Comprehensively Deepening Reforms* further stated the aim to consolidate the basic pension and medical insurance systems for urban and rural residents. Unifying the basic medical insurance systems for urban and rural residents is a long-term, complex task. Top-down design should be combined with a trial-and-error approach to unify these systems in a phase-by-phase manner. Overall, this goal can be achieved in three steps. First, the current basic medical insurance system for urban residents should be consolidated with the NRCMCS into a single basic medical insurance system for urban and rural residents. Second, free medical care services for employees of government departments/agencies and public institutions should be gradually phased out and replaced with basic employee medical insurance that also covers rural migrant workers. This will establish a uniform basic medical insurance system for urban employees. Third, the basic medical insurance systems for urban and rural residents and urban employees should be consolidated into a single basic medical insurance system that covers both urban and rural areas. This will help achieve institutional fairness and public resources sharing between urban and rural residents. Such consolidation should be carried out according to local conditions. Dongguan City, for

Table 9.1 Comparison of the three basic medical insurance systems in China

Form		For urban employees	For urban residents	NRCMCS
Coverage requirements	Guiding document	*Decision of the State Council on Building a Basic Medical Insurance System for Urban Employees* (GF [1998] No. 44)	*Guiding Opinion of the State Council on Implementing Pilot Projects for a Basic Medical Insurance System for Urban Residents* (GF [2007] No. 20)	*Notice of the General Office of the State Council on Forwarding the Opinion of the Ministry of Health and Other Ministries on Building a New Rural Cooperative Medical Care System* (GBF [2003] No. 3)
	Target group	Employees of all organizations in urban areas, the self-employed, part-time employees not covered by basic employee medical insurance and other flexible employees	Students at elementary, middle and high schools and other non-working urban residents not covered by the basic medical insurance system for urban employees	Rural residents
	Compulsory	Yes	No	No
Fundraising	Employer	Both employers and employees pay premiums for basic medical insurance according to relevant regulations; flexible employees pay the premiums individually according to relevant regulations	Employers may subsidize employee premiums for family insurance	–
	Employee		Households pay the most in premiums. The local authority may determine the level of fundraising according to the local economic situation	Each person pays a 10-yuan premium per year in the initial stage; the rate may be higher in economically developed regions
	Government subsidy	No	Yes	240 yuan per capita in 2012

(continued)

Table 9.1 (continued)

Form		For urban employees	For urban residents	NRCMCS
Benefits	Share of payment by public fund	About 70% in 2011	About 60% in 2011	About 60% in 2011
	Maximum payment by public fund	Over 6 times as high as the local average wage/salary or resident income	About 6 times as high as the local average disposable resident income	6 times as high as the local average net farmer income
Management and operation	Manager and operator	Human resource and social security authorities and affiliated operators	Human resource and social security authorities and affiliated operators	Health administrations and affiliated operators
	Level of uniform planning	Already city level in some regions	Uniform planning is made at the county and city levels; uniform planning at the city level was achieved in 2011	Uniform planning is generally made at the county and city levels, or, if not possible, at the township and town levels in the initial stage

Source Lin (2013)

example, has simply consolidated the three basic medical insurance systems into a single system. The more conservative model in Tianjin City took the approach of consolidating two systems, the basic medical insurance system for urban residents and the NRCMCS (Lin 2013).

9.4.2 Building a Uniform Basic Pension Insurance System for Urban and Rural Residents

There are several basic pension insurance systems in China, agencies and for urban employees, one for urban residents, the new pension insurance system for rural residents, and institutional configurations for retirees offered by government departments/agencies and public institutions (see Table 9.2). The Chinese government has had institutional configurations for retirees of government departments/agencies and public institutions since the founding of the People's Republic of China. Payments for these retirees are completely funded by the government. China set up the basic pension insurance system for urban employees in 1986 in order to support the reform of state-owned enterprises. After more than 20 years of continuous improvement, this system now already covers various enterprises, the self-employed and flexible employees in urban areas. In 2012, for example, this system covered 304 million people. Each covered retiree received 1721 yuan per month on average. In comparison, the new pension insurance system for rural residents was not established until after 2007. It covers rural residents at or above the age of 16. The pension insurance system for urban residents was set up based on pilot projects implemented in 2011 and mainly targets urban residents who are jobless or have no stable jobs. In 2012, the pension insurance systems for urban and rural residents already covered 484 million people in total. Today, China's basic pension insurance systems have achieved full coverage overall. Every elderly person who meets retirement requirements can receive monthly payments at varying amounts. This in itself is a great achievement. That said, China's basic pension insurance systems are divided between urban and rural areas, and insurance benefits vary widely from group to group. There are big differences in the level of uniform planning between these insurance systems. These programs also place heavy financial burden on governments who at times struggle to make relevant payments. Unsustainability is a growing risk. There are big differences between urban and rural pension insurance systems in terms of providing coverage, fundraising and payment, sources of money, the levels of uniform planning, and benefits and eligibility, in particular (Shao and Liu 2013). From a fairness perspective, consolidating the different types of pension insurance into uniform basic pension insurance for urban and rural residents is a necessary direction for future reforms.

The State Council decided at an executive meeting on February 7, 2013 that the new rural pension insurance system and the pension insurance system for urban residents would be merged into a single basic pension insurance system for urban and rural residents. The State Council officially released the *Opinion on Building a*

Table 9.2 Comparison among urban and rural basic pension insurance systems in China

	Urban			Rural
Insurance system	Basic pension insurance for urban employees	Institutional configurations for retirees of governments and public institutions	Pension insurance for urban residents	New rural pension insurance
Way of coverage	Compulsory	Compulsory	Voluntary	Voluntary
Target	Employees in enterprises, the self-employed and flexible employees in urban areas	Civil servants and employees of public institutions	Urban residents who are jobless or have no stable jobs	Rural residents at or above the age of 16
Way of fundraising	Cash basis combined with full funding	Cash basis	Full funding	Full funding
Source of money	Employers, employees and government	Government	Residents and government subsidies	Residents, collectives and government
Way of payment	Determined premiums	Determined benefits	Determined premiums	Determined premiums
Payment and government subsidy	A proportion of wages. Base: The monthly average wage of an employee in the previous year; proportions: 8% for employees and 20% for employers; funding assurance from the government	All premiums paid by the government	Fixed value. Ten levels between 100 and 1000 yuan; government subsidy: Premium subsidy of at least 35 yuan per person per year; basic pension of 55 yuan per person per month	Fixed value. Five levels between 100 and 500 yuan; government subsidy of at least 35 yuan per person per year; basic pension of 55 yuan per person per month
Level of uniform planning	Provincial	National	Municipal	County

(continued)

Table 9.2 (continued)

	Urban			Rural
Pension payment level	Pension = basic pension + pension from personal account; basic pension = (the monthly average wage of in-service employees in the previous year + the retiree's index- and taxation-based monthly average wage) ÷ 2 × the total number of paying years (incl. years viewed as paying years) × 1%; pension from personal account = the cumulative total in personal account ÷ the planned number of benefits-paying months	A certain proportion of pre-retirement wage	Pension = basic pension + pension from personal account; basic pension = 55 yuan; pension from personal account = the cumulative total in personal account ÷ 139	Pension = basic pension + pension from personal account; basic pension = 55 yuan; pension from personal account = the cumulative total in personal account ÷ 139

Source Shao and Liu (2013)

Uniform Basic Pension Insurance System for Urban and Rural Residents (GF [2014] No. 8) in February 2014, stating that the merger of the rural pension insurance system and the pension insurance system for urban residents into a single system would interconnect with the basic pension insurance system for employees, and would be implemented across China by the end of the 12th Five-Year Plan period. It further stated that a fair, uniform, standardized basic pension insurance system for urban and rural residents would be built by 2020 under the principles of full coverage, a broadened basic safety net, elasticity and sustainability. This new basic pension insurance system for urban and rural residents would be funded by individual payments, collective support and government subsidies. The Central Government would provide full-value subsidies to central and western regions, and 50% subsidies to eastern regions, according to the basic pension standards. Today, fifteen Chinese provinces have established a uniform basic pension insurance system for urban and rural residents. Difficulties in merging the aforementioned two pension insurance systems by 2015 are not expected. From a long-term perspective, the Chinese government must also accelerate the reform of the basic pension insurance system for governments departments/agencies and public institutions. Rather than limit pension insurance in this way, the Chinese government should build a basic pension insurance system that covers not just these organizations, but also urban employees, urban residents and rural residents. This will duly and ultimately consolidate the three separate pension insurance systems into a uniform urban and rural basic pension insurance system characterized by uniform, national planning, enabling all Chinese people to fairly share basic pension benefits.

9.4.3 Building a Uniform Social Assistance System for Urban and Rural Areas

Social assistance is a basic public service provided by the government to those in need. It is part of the social security system and includes three types of assistance: regular, special and temporary assistance. China has aggressively pursued the building of urban and rural social assistance systems in recent years. As part of this assistance program, the Chinese government has established a minimum living guarantee and medical assistance for urban and rural residents, established the "five guarantee" support system for rural households, and has provided temporary assistance and other social assistance systems. China has traditionally had separate urban and rural social assistance systems with severely uneven distribution of relevant benefits as a result of the traditional urban-rural divide. Today, urban social assistance systems target and cover a large audience and offer various programs including the minimum living guarantee and assistance related to medical care, housing, education, employment, legal work, and beggar assistance. In contrast, rural social assistance systems target and cover a small audience and offer few programs including minimum living guarantees, "five guarantee" support for households, and disaster assistance. Implementation of other social assistance programs has progressed very slowly. Under social assistance systems and policies that differentiate between urban and rural areas, rural areas lag far behind urban areas in terms of standards, levels, financial investment and institutional configurations. In 2010, for example, the average payment for urban social assistance was 5091.90 yuan per capita, but only 452.92 yuan per capita in rural areas, a gap of more than a factor of ten (Zhu 2012). In 2012, the average standard and spending for the minimum living guarantee in rural areas was only 52.2 and 43.5% that of urban areas respectively (see Table 9.3); the average medical assistance per capita in rural areas was also much lower than in urban areas.

In order to accelerate the provision of basic social security services on an equal basis, the Chinese government must go beyond the urban and rural boundaries and institutional barriers and build a uniform set of social assistance systems that provides equal benefits to urban and rural areas. To do so, the government must first unify urban and rural social assistance policies. Uniform standards and policies should be implemented in urban and rural areas under a uniform institutional framework, with the exception of a few urban- or rural-specific assistance programs, such as the "five guarantee" support for rural households and urban beggar assistance. A single set of social assistance systems covering basic living allowance and medical care, education, housing, employment, and legal assistance should be established to provide needy urban and rural people with equal social assistance services. A uniform minimum living guarantee and medical assistance systems, in particular, should be built at a faster rate. Second, the Chinese government should unify the definitions of social assistance targets in urban and rural areas. Urban and rural people targeted by social assistance should be identified under the same standards, with situations specific to urban and rural areas duly considered. It is necessary to further improve the temporary assistance system by covering permanent residents with non-local *hukou* and non-local rural migrant workers. Third, the Chinese government should increase

Table 9.3 Urban versus rural areas in China: average minimum living guarantee and spending (in yuan/person/month)

Indicator		2006	2007	2008	2009	2010	2011	2012
Average standard for minimum living guarantee	Urban	169.6	182.4	205.3	227.8	251.2	287.6	330.1
	Rural	70.9	70.0	82.3	100.8	117.0	143.2	172.3
	Rural/urban (%)	41.8	38.4	40.1	44.2	46.6	49.8	52.2
Average spending for minimum living guarantee	Urban	83.6	102.7	143.7	172.0	189.0	240.3	239.1
	Rural	34.5	38.8	50.4	68.0	74.0	106.1	104.0
	Urban/rural (%)	41.3	37.8	35.1	39.5	39.2	44.2	43.5

Source China Civil Affairs' Statistical Yearbook (2013)

rural support. Both central and local governments should increase investment in rural social assistance, further implement and improve rural medical care, education, employment, housing, and legal assistance, and continuously improve the standards for and levels of social assistance in rural areas. This will facilitate the fair distribution of social assistance resources between urban and rural areas.

9.5 Improving the Urban Administrative Area Management System

Today, China's urban administrative area management system still faces challenges, including the continued suspension of city establishment and excessive administrative levels. The assignment of administrative powers based on a city's administrative level has caused a severe and unfair imbalance in inter-regional distribution of resources. This in turn is an obstacle to the development of cities that may have great growth potential but is classified at a lower administrative level. From a macroscopic perspective, the current management system characterized by the urban-rural divide[1] also has hindered the process of urban and rural integration as well as improvements in the quality of urbanization. In order to fuel new urbanization, it is necessary to improve the urban administrative area management system. The near-term focus should be on improving the administrative division and levels of cities. The long-term focus should be on achieving an administrative pattern where urban and rural areas are administered under the same institutional framework.

9.5.1 Improving Administrative Division

To meet the requirements for urban and rural economic development, in April 1986 the State Council approved and forwarded the Ministry of Civil Affairs (MCA) *Report on Adjusting the Standards for Setting Up Cities and Conditions for City-administered Counties* (GF [1986] No. 46). This was followed by another MCA report approved and forwarded by the State Council in May 1993 entitled *Report on Adjusting the Standards for Setting Up Cities* (GF [1993] No. 38. Under these policies, counties turned into cities at an astonishing rapid rate across China. The number of cities grew from 353 in 1986 to 668 in 1997. The increased number of cities provided important carriers for new urban residents but at the same time sparked excessive local attempts to establish cities across China. In 1997, the CPC Central Committee and the State Council jointly released the *Notice on Further Enhancing*

[1]The so-called "urban-rural divide" refers to the urban-rural dual structure of governance in aspects such as population registration/management, planning/construction, public services, social security, fiscal system and administration according to urban and rural population or area criteria. It has caused a series of problems (Lu 2000).

Land Administration and Effectively Protecting Farmland (ZF [1997] No. 11), which suspended approvals to turn counties into cities. The subsequent development of Chinese cities entered a new stage characterized by macroscopic regulation and consolidation—some prefectures were turned into cities, counties/cities turned into districts, townships and towns merged, or townships turned into towns. By the end of 2012, the number of Chinese cities dropped to 657, of which cities at the county level decreased to 368 in number. The continuous decrease in city count has been in contrast with faster urbanization in China since 1996. The suspension of city establishment has not only failed to suit the needs for urbanization but has also hindered the growth of economically developed towns. Today, there are a large number of Chinese towns with a town-area population of over 200,000, and even over 500,000 people. Although these towns have exceeded the size thresholds for medium-sized or even large cities, these towns are forced to implement town-level planning and administration systems, which are far from suiting their growth needs.

It is therefore necessary to expeditiously resume the establishment of cities by turning eligible counties or towns into cities. This will help establish an urban system with Chinese characteristics and an administrative area system that satisfies the requirements for social development. To do so, several steps are needed. First, the Chinese government should make new standards for setting up cities. The present population and land sizes of built-up areas, population density and urban area development (e.g., the status of public infrastructures) should be used as the key indicators for making or improving city standards. Economic indicators such as the development of non-agricultural industries should be used as supporting indicators, especially for standards related to the transition of towns or counties into cities. The standards for setting up cities should be relatively low for western border and ethnic towns with smaller population sizes and lower levels of economic development but with important strategic positions. Second, the Chinese government should promote the transformation of counties into cities or districts depending on local conditions. Such upgrading should be based upon whether it will encourage local governments to work harder for local development and whether it will fuel economic development in relevant towns. Economically developed counties that have high urbanization rates and that have close economic ties to neighboring prefecture or higher-level cities should be turned into districts when appropriate. Turning cities at the county level into city-administered areas must be prohibited so as to prevent urban sprawl by larger cities. Third the Chinese government should upgrade towns into cities in a prudent manner. Chinese provinces such as Zhejiang, Guangdong and Shandong have started pilot projects intended to enhance the administrative powers of economically developed towns. The experience from these provinces should guide the prudent upgrading of towns into cities. Turning counties into cities differs from turning towns into cities in that the former does not involve any change in administrative level, while the latter will lead to a rise in administrative level. A town will have the administrative powers of a city at the county level after it is upgraded to a city.

9.5.2 Simplifying Administrative Levels

China has the world's largest number of administrative levels. Shortly after it began reform and opening up, China launched a system in which cities administered counties in order to allow central cities play a more important role. In March 1983, the State Council embraced the experiences of provinces such as Liaoning in implementing this city administration system and approved similar pilot projects in Jiangsu Province. This system was then implemented across China. By implementing a system in which counties are administered by cities, China expanded its administrative levels to include central, provincial (including autonomous regions and municipalities directly under the Central Government), prefecture (including autonomous prefectures and leagues), county/city/district and township/town administrative levels. This system has played a positive role in achieving economies of agglomeration and establishing closer urban-rural ties. However, with resource allocation biased toward administrative centers, central cities at the prefecture level have tended to take advantage of their administrative strengths to "absorb" people, industries and resources from administered counties on a large scale. This practice has deprived counties of growth opportunities and hampered expected ripple effects. Today, it is a common phenomenon in many prefecture level Chinese cities that human resources, funds and land quotas have been siphoned away from the counties to which they were allocated to the central urban areas of the administering city.

County-level administrative areas constitute an important level in China's administrative hierarchy. They play a significant role in socioeconomic development, especially in coordinated urban and rural development. The Chinese government should carefully analyze local experience before increasing the administrative powers of county-level governments. To facilitate socioeconomic development, fiscal power should be commensurate with administrative power, and powers should be commensurate with responsibilities. The Chinese government should establish systems, including fiscal systems, where provinces directly administer their respective counties/cities. In the meantime, important towns should be granted county-level administrative powers with regard to certain aspects of socioeconomic development in order to suit the needs of coordinated urban and rural development. Public management functions in fields such as construction, planning, road administration, sanitation and environmental protection should be gradually extended to townships and towns.

9.5.3 Achieving Urban and Rural Administration Under the Same Institutional Framework

Under the traditional separate urban and rural administration systems, rural issues are handled mainly by rural work offices, agricultural committees, poverty alleviation offices as well as agricultural authorities that deal with agriculture, stockbreeding, forestry and water resource issues. The other Chinese authorities typically focus on urban areas. In brief, the Chinese government has clearly prioritized urban areas over rural areas. Such separate urban and rural administration systems are a severe obstacle to urbanization and urban and rural integration, nor are they conducive to improving the quality of urbanization. In recent years, many local governments have set up offices to effectively coordinate urban and rural development and improve interdepartmental relations. Nonetheless, well-coordinated urban and rural development involves various aspects, making it impossible for any single government department to address it. From a long-term perspective, the concept of well-coordinated urban and rural development should be embodied in every facet of socioeconomic development. Functions specific to this concept should be included into the conventional functions of government departments/agencies at all levels. Long-term administrative mechanisms that cover both urban and rural areas should be created for planning, service and managerial purposes, in order to gradually achieve urban and rural administration under the same institutional framework. To this end, it is necessary to cease the prioritization of urban areas over rural areas, remove institutional barriers that have caused the urban-rural and even town-village divide, and cover rural areas with managerial functions of government departments/agencies at all levels. With respect to this last point, it is particularly important to cover rural areas with functions relevant to socioeconomic management and public services in fields such as planning, land administration, transport, communications, education, cultural development, sanitation, disease prevention and control, water services, safety/security and fire protection. Uniform planning, construction, protection and management should be implemented for urban and rural areas. Urban and rural policies also should be uniform.

The Chinese government should establish an integrated set of administrative systems characterized by powers that are commensurate to responsibilities, reasonable division of work, scientifically sound decisions, smooth implementation and effective supervision. In the process of urban and rural integration, future policies should treat urban and rural areas equally, with limited exceptions for policies directly related to land contracting, collective economy, agricultural production or urban development. Overall, standards, policies, management and services for urban and rural areas should all be uniform and equal.

References

Cai, F., Du, Y., & Wang, M. Y. (2003). *Political economy of labor mobility*. Shanghai: Shang People's Publishing House.

Cai, F., & Wang, M. F. (2009). The financial crisis' impact on employment and recommended countermeasures. In: *Analysis on the prospect of China's economy*. Beijing: Social Sciences Academic Press (China).

Hu, J. G. (2007). On the wage gap in China's urban labor market: Discrimination against people with rural *hukou*. *Statistics and Decision, 1,* 107–108.

Hu, X. Y. (2012). Accelerate the provision of uniform social security services for urban and rural areas. *Social Security Studies, 1,* 1–3.

Jiang, D. M. (2013). Build a single market for land used for urban and rural development. *People's Daily*, December 19, 2013.
Lin, S. (2013). Practices and thoughts on consolidating China's urban and rural medical insurance systems. *Journal of Changchun University, 1,* 1–5.
Lu, X. Y. (2000). How to go beyond policies that cause urban-rural split. *Dushu, 5,* 3–9.
Ministry of civil affairs of the People's Republic of China (2013). *China Civil Affairs' Statistical Yearbook (2013).* Beijing: China Statistics Press.
National Bureau of Statistics. (2013). *Migrant workers monitoring survey report 2012.* May 27, 2013.
Shao, W. J., & Liu, X. M. (2013). On how to consolidate the pension insurance systems for urban and rural residents in China. *Journal of Changchun University, 11,* 1369–1374.
Wang, X. D. (2012). On the reforming of China's fiscal system for uniform social security services in urban and rural areas. *Macroeconomic Management, 12,* 56–58.
Wang, Y. (2013). Establish an equitable employment system for urban and rural workers more rapidly. *Macroeconomic Management, 10,* 38–39.
Wu, Y., & Wu, G. (2009). The "213" vision about building a uniform land administration system for urban and rural areas. *China Real Estate Business*, July 27, 2009.
Yang, Y. Y., & Gu, Y. (2011). Use the surveyed urban unemployment rate as a binding indicator in the outline of 12th Five-Year Plan. *China Development Observation, 3,* 4–6.
Zhu, C. B. (2012). Inclusive growth and the integration of urban and rural social assistance systems: From the perspective of equal opportunities. *Seeker, 12,* 197–198.

MIX
Papier aus verantwortungsvollen Quellen
Paper from responsible sources
FSC® C105338

If you have any concerns about our products,
you can contact us on
ProductSafety@springernature.com

In case Publisher is established outside the EU,
the EU authorized representative is:
Springer Nature Customer Service Center GmbH
Europaplatz 3, 69115 Heidelberg, Germany

Printed by Libri Plureos GmbH
in Hamburg, Germany